T0091634

Glasses and Glass Ceramics for Medical Applications

Emad El-Meliegy · Richard van Noort

Glasses and Glass Ceramics for Medical Applications

Emad El-Meliegy
Department of Biomaterials
National Research centre
Dokki Cairo, Egypt
emadmeliegy@hotmail.com

Richard van Noort
Department of Adult Dental Care
School of Clinical Dentistry
Sheffield University
Claremont Crescent
Sheffield, UK
r.vannoort@sheffield.ac.uk

ISBN 978-1-4614-1227-4 e-ISBN 978-1-4614-1228-1
DOI 10.1007/978-1-4614-1228-1
Springer New York Dordrecht Heidelberg London

Library of Congress Control Number: 2011939570

Printed on acid-free paper

Springer is part of Springer Science+Business Media (www.springer.com)

Preface

Glass-ceramics are a special group of materials whereby a base glass can crystallize under carefully controlled conditions. Glass-ceramics consist of at least one crystalline phase dispersed in at least one glassy phase created through the controlled crystallization of a base glass. Examples of glass-ceramics include the machinable glass-ceramics resulting from mica crystallization, the low thermal expansion glass-ceramics resulting from β-eucryptite and β-spodumene crystallization, high toughness glass-ceramics resulting from enstatite crystallization, high mechanical strength resulting from canasite crystallization or the high chemical resistance glass-ceramic resulting from mullite crystallization.

These materials can provide a wide range of surprising combinations of physical and mechanical properties as they are able to embrace a combination of the unique properties of sintered ceramics and the distinctive characteristics of glasses. The properties of glass-ceramics principally depend on the characteristics of the finely dispersed crystalline phases and the residual glassy phases, which can be controlled by the composition of the base glass, the content and type of mineralizers and heat treatment schedules. By precipitating crystal phases within the base glasses, exceptional novel characteristics can be achieved and/or other properties can be improved.

In this way, a limitless variety of glass-ceramics can be prepared with various combinations of different crystalline and residual glassy phases. With the appropriate knowledge on the right way to modify the chemical compositions and the heat treatment schedules, one can effectively control the phase contents, scale the developed properties and control the final product qualities. Consequently, a skilled glass-ceramist is able to play with the constituting chemical elements and their contents in the composition to regulate the different ceramic properties.

Admittedly, the success in controlling functional properties is much more difficult if opposing properties such as high hardness and good machineability are desired. Similarly, achieving good chemical resistance in the presence of high content of alkalis and alkaline earths or rendering inactive glass ceramics into bioactive glass ceramics through composition modification are difficult to reconcile. Thus there are some real challenges and some serious limitations to what can be achieved.

This book includes five parts. The first part provides the context in which the classification and selection criteria of glass and glass ceramics for medical and dental applications are observed. This part starts with an introduction to medical glasses and glass ceramics, their classification and the specific criteria for various applications in order to show the clinical context in which the materials are being asked to perform. The grouping and arrangements of ions in silicate based glasses and glass ceramics are considered.

The second part deals with the manufacturing, design and formulation of medical glasses and provides a detailed description of theoretical and practical aspects of the preparation and properties of glasses. This part explains theoretically and practically how it is possible to predict final glass properties such as density, thermal expansion coefficient and refractive index from the starting chemical compositions. Next this part focuses on the manufacturing of the glasses and shows how to calculate and formulate the glass batches, melt, and cast glasses. The part also explains how to predict the right annealing point, transition point, and glass softening temperature of the base glasses.

The third part presents the manufacturing and methodology, the assessment of physical and chemical properties and the development of colour and fluorescence in medical glasses and glass ceramics. In addition, the microstructural optimization which is responsible for most of the valuable ceramic properties is considered. This part also explains how to optimize the microstructure so as to reach a uniform microcrystalline glass ceramic microstructure and gives examples of practical optimization such as mica and leucite-mica glass ceramics. The last chapter of this part deals with the selection of the glass compositions such that the materials can develop the correct colour and have the desired fluorescence. It also provides the ways for the development of colours and florescence in UV and visible light regions and a reliable quantitative measurement of colour and fluorescence in dental glasses and glass ceramics.

The fourth part presents a detailed description of the most prevalent clinically used examples of dental glass ceramics namely; leucite, mica and lithium disilicate glass ceramics, together with the encountered scientific and technical problems. This part explores in details the chemical composition, developed crystalline phases and the criteria for choosing the right chemical composition for different applications as veneering ceramics for coating metal alloys and glass ceramics for CAD/CAM applications. Appropriate solutions for common scientific and technical problems encountered with their industry and applications are discussed. The part also explores how to control and modify the chemical, thermal, mechanical, optical and microstructural properties of glass ceramic systems.

The fifth part provides a brief description of the chemical compositions, bioactivity and properties of bioactive glasses and glass ceramics for medical applications. This part also discusses different models of bioactive glass ceramics such as apatite, apatite–wollastonite, apatite–fluorophlogopite, apatite–mullite, potassium fluorrichterite and fluorcanasite glass ceramics.

The primary function of this book is to provide anybody with an interest in medical and dental glasses and glass ceramics with the wherewithal to start making their own glasses and glass-ceramics. Even if that is not their ambition then this book provides the reader with a greater understanding of the delicate interplay between the various factors that control the final properties of medical and dental glasses and glass-ceramics. This book is a valuable source of information for scientists, clinicians, engineers, ceramists, glazers, dental research students and dental technicians in the field of glasses and glass ceramics, and appeals to various other related medical and industrial applications.

Sheffield, UK

Emad El-Meliegy
Richard van Noort

Contents

Part V Bioactive Glass and Bioactive Glass Ceramics

List of Tables

List of Figures

Part I
Introduction to Medical Ceramics

Chapter 1
History, Market and Classification of Bioceramics

1.1 Bioceramics

A wide range of materials is used in the construction of medical devices and each material will interact in some way with the biological environment. These materials are generally described as biomaterials. A biomaterial is a synthetic material to be used in intimate contact with living tissue. A more precise definition of a biomaterial was given in 1986, at the Consensus Conference of the European Society for Biomaterials, when a biomaterial was defined as "a nonviable material used in a medical device, intended to interact with biological systems."

The field of biomaterials has grown and extended in its capacity to involve numerous scientific disciplines, including but certainly not limited to chemistry, geochemistry, mineralogy, physics, engineering, biology, biotechnology, human genetics, and medicine and dentistry. Despite rapid developments in such areas as tissue engineering, most biomaterials are still synthetic and used as implants to substitute for diseased or damaged tissues. Biomaterials cover a broad spectrum of materials including natural or synthetic, inorganic or organic, metals, polymers, or ceramics as shown in Fig. 1.1.

The main use by far is in the replacement of the hard tissues of the body such as knee and hip joint prostheses and perhaps the most extensive use of biomaterials is in the replacement of the oral hard tissues, namely enamel and dentine. Although many different kinds of biomaterials have been developed during the last two decades, they still need much control over the functional properties. For example, materials can degrade severely in demanding situations such as joint replacements due to wear and corrosion (e.g., polymers can wear out at a rate of 0.1–0.2 mm/year, while metals may corrode at a rate of 0.05 μm/year).

Ceramics that are used for reconstructive purposes as a bone substitute are termed "Bioceramics." Bioceramics is the branch of biomaterials that represents around 50% of the world consumption of biomaterials. Bioceramics are needed to alleviate pain and restore functions to diseased or damaged parts of the body. A significant contributor to the need of bioceramics is that bone is especially susceptible to

E. El-Meliegy and R. van Noort, *Glasses and Glass Ceramics for Medical Applications*,
DOI 10.1007/978-1-4614-1228-1_1,

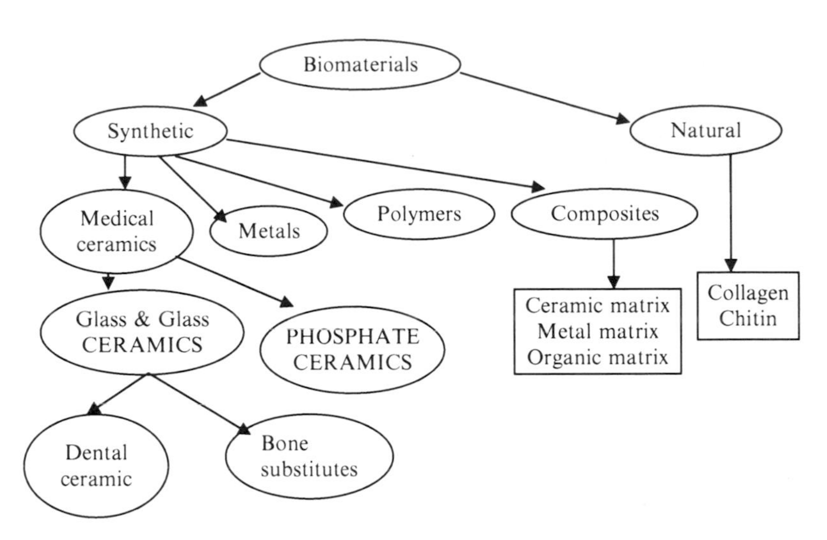

Fig. 1.1 Classification of biomaterials

fracture in older people, because of a loss of bone density and strength with age. Bone density decreases because bone-growing cells (osteoblasts) become progressively less productive in making new bone and repairing micro-fractures. The lower bone density greatly deteriorates the strength of the porous bone existing in the ends of long bones and in vertebrae.

Although widely used as a material for skeletal reconstruction, the most likely place where patients will be exposed to bioceramics is when they need dental treatment. Teeth are made from enamel and dentine and these tissues, unlike bone, do not have the capacity to repair if they get damaged due to dental diseases such as caries and periodontal diseases. Millions of people seek dental treatment every year and the demand for esthetic tooth like restorations is increasing. Ceramics are well suited to meeting this demand and dental materials represent one of the fastest growing applications of bioceramics. Ceramics are used in a range of dental filling material such as glass fillers in glass ionomer cements and resin composites and are also extensively used in the construction of crowns and bridges to restore or replace missing teeth.

The great challenge facing the use of bioceramics in the body is to replace diseased hard tissues such as bone, dentine, and enamel with a material that can function for the remaining years of the patient's life. Because the life span of humans can now exceed 80+ years and the need for spare parts begins at about 60 years of age or even earlier in the case of dental treatment, bioceramics need to last for many decades. The excellent performance and long-term survival of well-designed bioceramic prostheses in these demanding clinical conditions represents one of the greatest challenges of ceramics research.

Bioceramics are mainly based on the preparation and use of synthetic mineral phases with controlled properties. So we can state categorically that bioceramics are synthetic mineral phases and their properties will be typically synthetic phase

Table 1.1 Clinical applications of bioceramics

Application	Ceramic materials
Orthopedic load bearing	Alumina, partially stabilized zirconia
Dental orthopedic	Bioactive glasses, glass ceramics, alumina, partially stabilized zirconia
Dental implants	Alumina, hydroxyapatite, bioactive glasses
Temporary bone space fillers	Tricalcium phosphate
Alveolar ridge	Bioactive glass ceramics, alumina
Spinal surgery	Bioactive glass ceramics, hydroxyapatite
Maxillofacial reconstruction	Bioactive glasses, glass ceramics

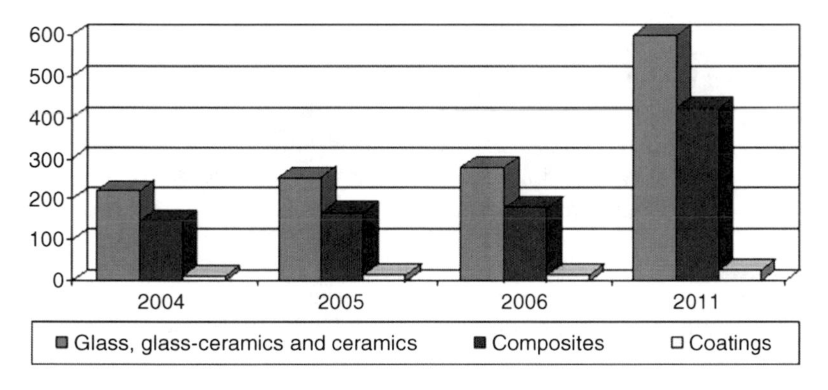

Fig. 1.2 Market of glass ceramics, composites, and coatings

properties. As we enter the twenty-first century, the field of bioceramics is becoming one of the most academically exciting areas of materials science and engineering. Early activities in this field (50 years ago) dealt with the selection of well-established synthetic materials to fabricate implants for use in medicine and dentistry (see Table 1.1).

The worldwide global market for bioceramics increased from $377.7 million in 2004 to $431.4 million in 2005 with sales reaching an estimated $473.9 million by the end of 2006. The sale of glasses, glass–ceramics, and ceramics represents the largest share of this market. At an average annual growth rate of 17.2%, this market is expected to exceed $1.0 billion by 2011. Glass filled composites show the highest growth rate through the forecast period with an average annual growth rate of 18.7% reaching $423.3 million by 2011 as indicated in Fig. 1.2. This process of finding new uses for industrially engineered ceramics will undoubtedly continue, one example being the developments in using zirconia in dentistry to produce crowns and bridges. However, there is a growing trend toward biomaterials that are designed to produce a well-defined interaction with the biological environment, one obvious example being the bioactive glasses, which stimulate the formation of new bone.

1.2 Classification of Bioceramics

The classification of any group of material is invariably a source of contention as no classification can ever be ideal. When considering bioceramics a classification based on the interaction with the biological environment would seem to be the most appropriate. When a bioceramic is implanted within the human body, tissue reacts toward the implant in a variety of ways depending on the material type. It has been accepted that no foreign material placed within a living body is completely compatible. The only substances that conform completely are those manufactured by the body itself. No synthetic material can be considered as being inert as all materials will produce some sort of a response from living tissue. These materials will be recognized as foreign and may initiate any of a range of tissue responses. The mechanism of tissue interaction depends on the tissue response to the implant surface. Four types of bioceramic-tissue interactions are shown in Table 1.2.

The challenge for the materials scientists is to develop new ceramics that produce the most appropriate response that the clinical situation demands. In general, bioceramics can be classified according to their tissue response as being passive, bioactive, or resorbable ceramics.

1.2.1 Biopassive (Bioinert) and Bioactive Materials

Historically, the function of biomaterials has been to replace diseased or damaged tissues. First generation biomaterials were selected to be as *biopassive* as possible and thereby minimize formation of fibrous tissue at the interface with host tissues. Biomaterials that initiate a host response risk damaging adjacent tissues, resulting in fibrous encapsulation of the implant or in more toxic situations may lead to worse sequlae, including necrosis and sequestration of the implant. Fibrous encapsulation is a host defense mechanism, which attempts to isolate the implant from the host. This usually is a response to mildly irritant biomaterials. Where an implant is in close proximity to bone and if there is a lack of a fibrous tissue layer, that is, a close apposition of new bone to the surface of the implant, this is described as osseointegration.

Osseointegration. This term was first described by Brånemark and coworkers and defined as direct contact, at the light microscope level, between living bone and the implant. Dorland's Illustrated Medical Dictionary defines osseointegration as "the

Table 1.2 Tissue responses to bioceramics

1. If the material is toxic, the surrounding tissue dies
2. If the material is nontoxic and biologically passive, it is encapsulated with fibrous tissue or bone
3. If the material is nontoxic and bioactive, an interfacial bond forms
4. If the material is nontoxic and dissolves, the surrounding tissue replaces it

direct anchorage of an implant by the formation of bony tissue around the implant without the growth of fibrous tissue at the bone-implant interface."

Biomaterials used for bone repair can be further classified as osteoconductive and osteoinductive.

Osteoconduction. An osteoconductive surface is one that permits bone growth on its surface or down into pores, channels, or grooves such that it conforms closely to a material's surface. Osteoconductive materials provide an interfacial surface that permits bone migration eliciting a response along the implant/tissue interface. Typically, biomaterials containing an apatite phase are osteoconductive. Glass–ceramics containing an apatite phase can potentially initiate an osteoconductive bone response but this is dependent of the chemical composition and surface texture of the material.

Osteoinduction. This term means that primitive, undifferentiated and pluripotent cells are somehow stimulated to develop into the bone-forming cell lineage. One proposed definition is "the process by which osteogenesis is induced." Osteoinductive materials stimulate local osteogenic stem cells or osteoprogenitor cells to proliferate within close proximity of the implanted material and deposit new bone, independent of the site of implantation.

An even more desirable interfacial response is when a bond forms across the interface between implant and tissue. Such a biomaterial is referred to as *bioactive*. Bioactivity allows the biomaterial to form a strong bond with bone leading the greater long-term stability. This type of interfacial reaction requires the material to have a controlled chemical reactivity, most commonly in the form of controlled dissolution. Ideally, the biomaterial should dissolve at a controlled rate that matches the rate of new tissue deposition leading to a state of dynamic equilibrium. The breakdown products of the biomaterial are degraded into excretable components or are digested by macrophages. Calcium phosphate ceramics and bioactive glasses are able to produce interfacial bonding capabilities with host tissues, leading to the concept of bioactive materials.

1.3 Mechanisms of Bioactivity

Bone deposition on an implant surface is, in some ways, similar to the process within living tissue. The early stages involve the deposition of osteopontin, which is a protein responsible for osteoblast adhesion on an implant surface. The implant surface is then colonized by osteoblastic progenitor cells, which differentiate into osteoblasts and in turn secrete a collagenous matrix. Mineral formation within the matrix is then mediated by osteocalcin and osteopontin and leads to the formation of mature bone. The implant surface, in turn, may form a biologically active hydroxycarbonate apatite (HCA) layer, which is responsible for interfacial bonding. This surface HCA layer can be formed by the following mechanisms.

1.3.1 Formation of a Silica-Rich Surface Layer

This theory was first promoted by Hench to explain the behavior of bioactive glasses that are based on compositions containing specific quantities of P_2O_5, SiO_2, Na_2O, and CaO. The bioactivity is a function of the chemical reactions between ions leached by the glass/glass–ceramic and ions present within the surrounding biological environment. Surface porosities on the bioactive glass may provide some mechanical interlocking aiding the bond. It is possible that there may even be a genetic engagement of the local osteoblasts that govern the biological response of the host tissue to be bioactive. In order for new bone to form it is necessary for osteoprogenitor cells to undergo mitosis, requiring the correct chemical stimuli from their local environment to instruct them to enter the active segments of the cell cycle. The formation of a surface HCA layer is believed to be beneficial but not critical for the provision of this chemical stimulus. The key phenomenon is the controlled rate of release of ions, particularly appropriate concentrations of soluble silicon and calcium ions.

1.3.2 Direct Precipitation of Apatite

The presence of an apatite containing crystalline phase within the implanted biomaterial has been postulated to act as a nucleation site for further apatite deposition leading to an interfacial bond with living bone. This theory may be used to explain interfacial bonding between an implant and bone, in the absence of a silica-rich surface layer.

1.3.3 Protein Mediation

The adsorption of protein to biomaterial surfaces is known as the "Vroman effect" and considerable research has been conducted on the interaction of serum proteins with a variety of surfaces. It has been suggested that the adsorption of serum proteins to an implant surface may influence osseointegration. Although the role of serum proteins on the surface of titania implants has been demonstrated, the extent to which protein adsorption is a key determinant of osseointegration or bioactivity in calcium phosphate ceramics, bioactive glasses, and glass–ceramics is less certain, and in some circumstances may even interfere with the process.

1.4 Biopassive Ceramics

Bioceramic that approach most closely the concept of being biopassive are high-purity dense zirconia and alumina. When such a material is described as being biopassive it shows a distinct lack of reaction with the biological environment.

Fig. 1.3 Soft tissue encapsulation of a zirconia implant

The nature of the reaction of the biological environment depends on the local circumstances. If the implant is placed in close proximity to bone and left undisturbed, the bone will fill the space between the implant and the old bone with newly formed bone, resulting in close apposition of new bone to the implant surface. However, if there is excessive mobility of the implant then it is more likely that a fibrous capsule will form.

Al_2O_3 has been used in orthopedic surgery for more than 20 years in total hip prostheses, because of its exceptionally low coefficient of friction and minimal wear rates. It is also used in dental implants, because of its combination of excellent corrosion resistance, good biocompatibility, low friction, high wear resistance, and high strength. As is often the case these materials are not pure and in the case of alumina a very small amount of MgO ($<0.5\%$) is used as a sintering aid to limit grain growth during sintering.

Another material that evokes a similar response to alumina is zirconia (ZrO_2), which tends to show fibrous encapsulation when implanted in bone (Fig. 1.3). Zirconia has come to prominence recently, particularly in dentistry, due its combination of excellent strength and toughness. This has been achieved by the process of transformation toughening, whereby a small addition of another element such as yttrium prevents the zirconia from transforming on cooling from its tetragonal crystalline state to its more stable monoclinic state at room temperature, hence being referred to as partially stabilized zirconia (PSZ).

Under an externally applied load the stress generated transforms the tetragonal zirconia to its monoclinic form and this is accompanied by an expansion in volume. This generates a compressive stress in the structure, which counteracts the tensile stresses, especially those at a crack tip and prevents the crack from advancing. Some of the properties one might expect from these two ceramics are presented in Table 1.3.

Table 1.3 The properties of the most common passive ceramics (after L. Hench 1991)

Properties	High alumina ceramics	PSZ
Content (wt.%)	>99.5	>97
Density (g/cm^3)	3.93	6
Average grain size (μm)	3–6	1
Surface roughness, Ra (μm)	0.02	0.008
Hardness (Vickers. HV)	2,300	1,300
Compressive strength (MPa)	4,500	–
Bending strength (MPa)	550	1,200
Young's modulus (GPa)	380	200
Fracture toughness (K_{ic}) MPa. m$^{1/2}$	5–6	15

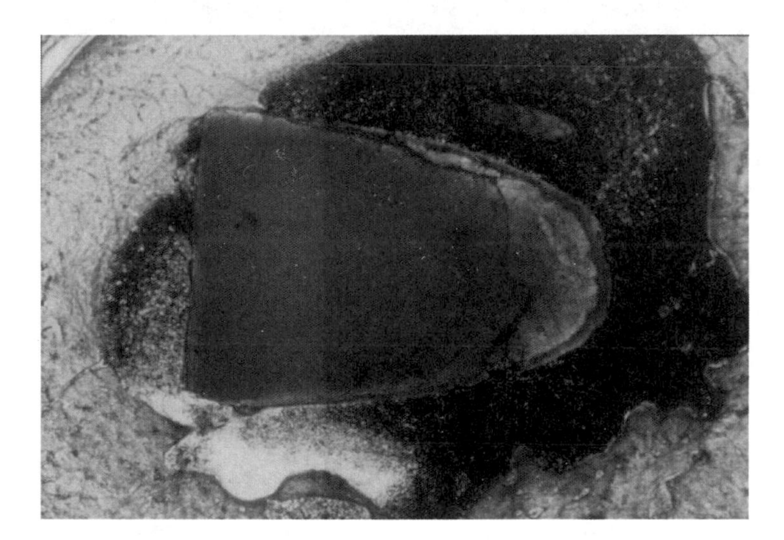

Fig. 1.4 New bone formation around a hydroxyapatite implant after as little as 4 weeks

1.5 Bioactive Ceramics

Bioactive ceramics refers to materials, which upon being implanted within the human body interact with the surrounding bone in such a way as to encourage the formation of new bone as well as forming an interfacial bond with the new bone being laid down. An example of a bioactive ceramic is synthetic hydroxyapatite, which encourages the formation of new bone on the surface of the implant, as shown in Fig. 1.4.

Bioactive ceramics have been described as osteoconductive, meaning that these facilitate the formation of new bone structure. This contrasts with passive ceramics where one may see close apposition of new bone to the implant surface but this is in no way enhanced by the presence of the implant and there is no formation of a chemical bond between the surface of the implant and the new bone. However, the

Table 1.4 Types of tissue attachment for bioactive ceramics

Bioceramic-tissue attachment	Bioceramics
Porous implants attachment by bone in-growth (biological fixation)	Porous hydroxyapatite
Attach by chemical bonding with the bone (bioactive fixation)	Bioactive glasses and glass–ceramics

time dependence, the strength, the thickness, and the mechanism of attachment will differ for different ceramic materials. The different types of tissue attachment that can be induced, depending on the bioceramic used, are shown in Table 1.4.

In those situations where tissue attachment is achieved by bone in-growth into surface pores, for the tissue to remain viable and healthy, pores must be >100–150 μm in diameter. Details of current clinically used bioactive ceramics based on the type of their mineral phases are summarized in Table 1.5.

1.6 Resorbable Bioceramics

Resorbable bioceramics are designed to degrade gradually in the biological environment, at the same being replaced with new tissue, generally speaking this will be new bone. In this case the interface is a hive of activity, with new bone being laid down as the bioceramic dissolves. This situation is akin to the natural situation where old bone is being replaced by new bone. A key feature of this interaction is to maintain the strength and stability of the interface while the bioceramic is being replaced with new bone. Thus the rate of resorption of the bioceramic has to match the formation of new bone, which is very difficult to achieve. It is also important that the degradation products of the bioceramic can be readily metabolized without causing any local or systemic adverse reaction. An example of a bioceramic that has met with a certain degree of success is tricalcium phosphate (TCP), so long as the demands on strength are not too high.

1.7 Currently Used Glasses and Glass Ceramics

1.7.1 Bioactive Glasses

Bioactive glasses are a group of surface reactive glasses that release ions into the local environment, which can then trigger a range of biological responses. The most desirable response is for the glass to stimulate the formation of new bone by the release of sodium, calcium, and phosphate ions. This group of materials was first developed in the late 1960s by Larry Hench and colleagues at the University of Florida. The first bioactive glass to be commercialized was Bioglass®, also known

Table 1.5 Composition and properties of a range of bioactive ceramics, after L. Hench (1998)

Property	Bioglass 4555	S45PZ	Glass–ceramic Ceravital	Glass–ceramic Cerabone A/W	Glass–ceramic Ilmarplant L1	Glass–ceramic Bioverit	Sintered. Hydroxyapatite Ca10(PO_4)6 $(OH)_2$ (>99.2%)	Sintered β-3 $CaO{\cdot}P_2O_5$ (>99.7%)
Composition (wt%)								
Na_2O	24.5	24	5–10	0	4.6	3–8		
K_2O	0		0.5–3.0	0	0.2	3–8		
MgO	0		2.5–5.0	4.6	2.8	2–21		
CaO	24.5	22	30–35	44.7	31.9	10–34		
Al_2O_3	0		0	0	0	8–15		
SiO_2	45.0	45	40–80	34.0	44.3	19–54		
P_2O_5	6.0	7	10–50	16.2	11.2	2–10		
CaF_2	0			0.5	5.0	3–23		
B_2O_3	0	2						
Phase[l]	Glass	Glass	Apatite Glass	Apatitie β-Wollastonite Glass	Apatite β-Wollastonite Glass	Apatite Phlogopite Glass	Apatite	Whitlockite
Density (g/cm^3)	2.6572			3.07		2.8	3.16	3.07
Vickers hardness (HV)	458 ± 9.4			680		500	600	
Compressive strength (MPa)			500	1,080		500	500–1,000	460–687
Bending strength (MPa)	42			215	160	100–160	115–200	140–154
Young's modulus (GPa)	35		100–150	218		70–88	80–110	33–90
Fracture toughness, K^{1C} (MPa m^{1,2})				2.0	2.5	0.5–1.0	1.0	
Stow crack growth, *n* (unitless)				33			12–27	

Table 1.6 Composition of different grades of Bioglass® [after Hench (1972)]

Component	SiO_2	CaO	Na_2O	P_2O_5	CaF_2
45S5 bioglass	45	24.5	24.5	6	–
45S5.4F bioglass	45	14.7	24.5	6	9.8
45B15S5 bioglass	30	24.5	24.5	6	–
52S4.6 bioglass	52	21	21	6	–
55S4.3 bioglass	55	19.5	19.5	6	–

as 45S5 glass. A range if compositions of bioactive glasses are presented in Table 1.6.

A key feature of Bioglass® is that it is low in SiO_2 and high in Na_2O and CaO contents and that it has a high CaO/P_2O_5 ratio, which makes Bioglass® highly reactive to aqueous media and gives it the characteristic if being bioactive. When a bioactive glass is immersed in a physiological environment an ion exchange process occurs in which modifier cations (mostly Na^+) in the glass exchange with the aqueous cation H_3O^+ in the external solution. This is followed by hydrolysis of the glass surface in which Si–O–Si bridges are broken, forming Si–OH silanol groups, and the glass network is disrupted. Subsequently, the silanols reform by a condensation reaction into a gel-like surface layer, which is depleted of sodium and calcium ions.

The next phase takes place in the environment immediately surrounding the implant in that an amorphous calcium phosphate layer is deposited on the gel, which eventually remineralizes into crystalline hydroxyapatite, mimicking the mineral phase that is naturally contained within vertebrate bones. The interest in bioactive glasses has increased year on year due to the effective control of its chemical, physical, and technological parameters. Bioactive glasses are used in a wide variety of applications, although mainly in the areas of bone repair and bone regeneration via tissue engineering. Applications include:

- Synthetic bone graft materials for orthopedic, craniofacial, maxillofacial, and periodontal repair
- Porous scaffolds for tissue engineering of bone
- Treating dentine hypersensitivity and promoting enamel remineralization

1.7.2 Glass–Ceramics

Glass–ceramics are made by the controlled crystallization of a glass and their properties are controlled by the mineral phases created. The gradual expansion in glass–ceramic technology has brought about new uses for many phases in the manufacture of a variety of industrial products. Almost all of the synthetic minerals are becoming products with a range of industrial applications. These minerals are synthesized in a way that is similar to what is happening in nature. Some mineral phases are synthesized using heat treatment or heat treatment together with pressure application, just like minerals with a hydrothermal origin. The difference is that the synthesized mineral phases are produced in a short time compared to the geological periods of formation.

Glass–ceramics are ideal materials for the design and manufacture of bioceramics with superior mechanical properties. It is very important to take note of the contribution of the predominant constituting mineral phase in bioceramics. Current bioactive glass ceramics include apatite glass ceramics, apatite–mullite glass ceramics, apatite–wollastonite glass ceramics, fluorcanasite glass ceramics, and potassium fluorrichterite glass ceramics. There is a close relationship between the constituting mineral phases and ceramic properties because almost all ceramic products have been manufactured from synthetic mineral phases. Also, the mineral phase and the glass composition have a profound influence on the biological response in determining if a bioceramic is passive, bioactive, or bioresorbable.

Glass–ceramics are generally fabricated from starting materials, which are naturally occurring minerals and rocks or their extracted reagent grade derivatives. The mineral phases constitute the majority of the medical ceramic microstructure and are considered the primary phases, while the matrix or the glassy phase is considered a secondary phase. The relations among the primary phases, the secondary phases, and the incorporated porosity are termed the ceramic microstructure, which is responsible for the developed physical, chemical, biological, and mechanical properties.

Therefore, there is an intimate relation between the microstructure, the developed phases, and the starting chemical compositions of glass ceramics. For example, the development of canasite or miserite increases the modulus of rupture, which can exceed 400 MPa and the fracture toughness can be as high as 3 MPa $m^{1/2}$. Also, the crystallization of beta-spodumene significantly hardens the glass ceramics by increasing the microhardness. On the other hand, the development of fluorophlogopite in some glass ceramics improves the machinability characteristic of this bioceramic. Also the chemical durability of bioceramics can be improved by the development of either mullite or wollastonite mineral phases in the glass ceramics. So it is possible to control the properties of bioceramics by controlling the type and content of the mineral phases.

Glass–ceramics based on the formation of an apatite crystalline phase are close to natural bone in their chemical composition, but the application of these materials in orthopedics is limited due to their lack of strength and toughness. At present many novel glass–ceramics are being explored, which are made up of multiple crystal phases including that of apatite. It is possible these new glass–ceramics will deliver the improvement of the strength sought along with the desired bioactivity and biocompatibility.

1.7.3 Dental Ceramics

Teeth consist of several components, primarily enamel, dentine, and pulp. If lost or damaged, a tooth cannot be repaired or regenerated. Restorative dentistry is concerned with the repair of damaged teeth and their supporting structures. A major contributor to the need of dental restorations is the progressive deterioration of teeth by dental caries and tooth loss. Dental caries and tooth loss are the most common diseases in developed countries and affect 60–90% of schoolchildren and the vast majority of adults. The enamel is found to dissolve by the attack of bacterial acid and appears as brownish or black discoloration of the teeth. On progression, this converts into cavities in the tooth that may reach the dentin. Dental caries commonly occurs on chewing surfaces or the interdental surfaces in posterior teeth. During this time people may complain of sensitivity to hot, cold, sweet, or sour drinks. At this stage the mineralized tissue that has been lost can be restored with dental restorations to restore health and esthetics. If left untreated this dental caries may lead ultimately to severe pain as any further lack of action to repair the situation will lead to involvement of the pulp and possibly subsequent abscess formation, which needs much more sophisticated treatment.

The history of dental ceramics can be traced back as far as ancient pharaoh times, where tooth replacement and prostheses were made from glass or ivory held in place with gold bands and wires. The Egyptians in 3000 BC numbered tooth doctors as medical specialists. The Ebers papyrus (http://en.wikipedia.org/wiki/Ebers_Papyrus) described established medical and surgical procedures used for dental disorders. Porcelain for decoration was probably first made by the Chinese during the Tang dynasty (618–907), who developed the techniques for combining the proper ingredients and firing the mixture at extremely high temperatures as it evolved from the manufacture of stoneware. European manufacturers responded by trying to make hard porcelain themselves, but the subject remain a vague secret for a long time. Nevertheless, some of their experiments resulted in beautiful soft-paste porcelain. The first European soft porcelain was produced in Florence, Italy, about 1575. However, by the 1700s porcelain was being manufactured in many parts of Europe and was starting to compete with Chinese porcelain. France, Germany, Italy, and England became the major centers for European porcelain production.

The dental application of porcelain dates back to 1774, when a French apothecary named Alexis Duchateau considered the possibility of replacing his ivory dentures with porcelain. Ivory, being porous, soaks up oral fluids and eventually becomes badly stained, as well as being highly unhealthy. Duchateau, with the assistance of porcelain manufacturers at the Guerhard factory in Saint Germain-en-Laye, succeeded in making himself the first porcelain denture. Since then, other materials such as more recently polymethyl methacrylate has helped to replace porcelain for denture applications. However, porcelain teeth, in conjunction with a pink acrylic denture base, are still extensively used.

The beginning of industrial dental porcelain goes back to year 1837, when Stockton made the first ceramic teeth. The properties of dental porcelain were improved by introducing vacuum firing in the year 1949. The growing demand for tooth colored restorations led to improvements in ceramic formulation and firing techniques. Because of the limitations in the strength of existing porcelains, during the 1970s great improvements were made in the area of porcelain veneers supported by a metal framework. Although this technique proved very successful and is still

Table 1.7 Dental ceramics

Name	Application
Feldspars	Veneers for alumina based cores and resin-bonded laminates
Alumina reinforced feldspars	Core for anterior crowns
Leucite containing feldspars	Veneers for metal, zirconia substructures and as resin bonded laminate veneers
Leucite glass–ceramics	Resin-bonded laminate veneers, anterior crown, and posterior inlays
Fluormica glass–ceramics	Resin-bonded laminate veneers, anterior crown, and posterior inlays
Li-disilicate glass–ceramic	Core for anterior and posterior crowns and bridges
Fluorapatite glass–ceramics	Veneer for lithium disilicate core
Glass infiltrated spinel	Core for anterior crowns and bridges
Glass infiltrated alumina	Core for anterior and posterior crowns and anterior bridges
Pure alumina	Core for anterior and posterior crowns and bridges
Partially stabilized zirconia	Core for anterior and posterior crowns and bridges

widely used to this day, it has its limitations. The principal difficulties associated with porcelain fused to metal restorations were the need to match the coefficient of thermal expansion of the porcelain to that of the metal, in addition to the need for opaque porcelain to mask the color of the metal substructure.

The porcelain-fused-to-metal technique has made it possible to fabricate dental restorations with more complicated structures, such as porcelain jacket crowns and bridges. Nevertheless, the demands for ever better esthetics and a growing resistance to the use of metals in the mouth, due to an increasing incidence of metal related adverse reactions, has made all-porcelain restorations a desirable goal. In order to avoid the need for a metal substructure, numerous efforts have been directed at replacing the metal substructure with a high-strength ceramic substructure.

The earliest attempts to strengthen dental porcelain usually involved the inclusion of strengthening oxide particles in the base porcelain. Examples of strengthening by oxides include zirconium oxide and aluminum oxide. In more recent years we have seen the introduction of glass infiltrated ceramics and as a consequence of developments in CAD-CAM technology it has become possible to use pure alumina and PSZ.

The advent of procedures that allowed a translucent ceramics to be bonded to enamel using a combination of etching, silanes and resins, resulted in the use of thin veneering shells (~700 μm thick) of ceramic to be bonded to the visible surfaces of front teeth to mask discoloration or defects of the anterior teeth. The materials used for the construction of the veneers are either simple feldspathic glasses or leucite reinforced feldspathic glasses, where the latter can be formed as a glass–ceramic. In these situations high strength and fracture toughness of the ceramics appear to be less important criteria for clinical success than esthetic potential. These simple structures serve mainly as esthetic surfaces and are clinically quite successful, although they are often made of the weakest dental ceramic.

Crowns are more complex prostheses that completely replace all external tooth structure on single teeth. Crowns are essentially thin-walled (1,000–2,000 μm thick) full coverage shells that can be composed of dental ceramic fused to either metal or high-strength ceramic substructures or can be composed entirely of an esthetic dental ceramic. In the case of the latter, where strength and toughness become a significant requirement, such as in posterior crowns where the loads are much higher, a glass–ceramic based on lithium disilicate has been recently introduced. For more extensive restorations such as bridges the search is still on to find a glass–ceramic alternative to alumina or zirconia. Thus dental bioceramics encompass a wide variety of ceramic materials including a range of glass–ceramics (Table 1.7).

Chapter 2
Selection Criteria of Ceramics for Medical Applications

2.1 Biocompatibility

Biocompatibility relates to the actions of the biomaterials when brought in contact with the living tissues and how these materials are able to integrate and react with the surrounding tissue. Biocompatibility is defined as:

> "The ability of a material to function in a specific application in the presence of an appropriate host response"

The biocompatibility is the most important issue for the application of medical ceramics whether biopassive, bioactive, or resorbable ceramics. Medical ceramic materials often include glasses, glass ceramics, and ceramic–polymer bioactive composites. The ceramics may be manufactured either in porous or in dense form, in bulk or in granules or in the form of coating layers. The biocompatibility is considered to be one of the most beneficial factors in using synthetic ceramics innovations in medicine. Any new glasses and glass–ceramics developed for medical purposes should possess this extremely important property of biological compatibility with living tissue.

During the past three decades, there have been major advances in the development of medical ceramics for skeletal repair and reconstruction and the biological side effects of dental ceramics compared with other restorative materials is considered to be low. For example, restorative dental ceramics are generally considered to be the most biocompatible of all dental materials. They are not generally known to cause biological reactions, except for wear on the opposing dentition and/or restorations. All implanted dental devices must fulfill the requirement of being biocompatible and the bioactivity of dental ceramics will differ according to the proposed application.

A range of tests can be conducted to establish the biocompatibility of a biomaterial. These include in vitro test for cytoxicity, microbial test to determine whether or not the material inhibits or stimulate the growth of microorganisms and genotoxicity/mutagenicity and carcinogenicity tests to assess possible systemic

E. El-Meliegy and R. van Noort, *Glasses and Glass Ceramics for Medical Applications*,
DOI 10.1007/978-1-4614-1228-1_2,

reactions. There is also a range of in vivo test that can be conducted, which include animal implantation tests to assess the local tissue response based on histological assessments, systemic toxicity tests such as LD_{50} (lethal dose for 50% of the test animals after oral uptake), sensitization, and/or irritation tests or pulp tests to determine the specific local reactions when considered relevant.

The preclinical evaluation of the cytotoxicity of biomaterials is conducted according the methods described in ISO 7405:1997. These tests generally involve producing a monolayer of cells on the surface of the material and monitoring the cellular response. The relative toxicity of materials can be determined by measuring cell death using an Alamar Blue staining technique from which it is possible to calculate the LD_{50} level of exposure. Further research is in process to develop more sophisticated models such as 3D skin and oral mucosal models that can examine the histology and measure local inflammatory markers.

Another aspect of the biocompatibility of ceramics is the potential for a radioactive component to be incorporated as one uses a wide range of natural minerals. For example, some of the products used in the early 1990s raised concerns about radioactivity because of impurities such as uranium and thorium in the zirconium oxide powder. Uranium, thorium, and some of their decay products can be present as impurities in zirconia powders even if powder manufacturing processes provide for an efficient separation of such elements with a different concentration according to the level of purification of the material and of the powder manufacturing process.

2.2 Radioactivity

All glasses and ceramics to be used for medical and dental applications should be tested for the presence of radioactivity. This can be done by taking a 500-g sample of the as received of the product powder, milling it in a suitable ball mill and sieving to yield powder with particle size of less than 75 μm. A sample volume of 60 ml of the ceramic powder is placed in a scintillation counter to determine the activity concentration of uranium-238 by neutron activation. The activity concentration should be no more than 1.0 Bq/g of uranium-238 (ISO 13356).

2.3 Esthetics

Esthetics is a primary consideration in the use of ceramics to replace missing tooth structure. Today esthetics are of such paramount concern that the only medical materials that in any way provide a durable and satisfactory solution to the esthetic repair of teeth are ceramics. For example, dental porcelains attempt to meet the challenges to achieve an appropriate visual appearance, whilst providing sufficient strength to accept the loads placed upon teeth during function. The esthetics of dental ceramics are characterized by three optical properties, namely: color, translucency, and texture.

- *Color* – The color of an object our eye detects is a function of the light source, providing the spectrum of light (380–700 nm) hitting surface, how the object transforms this spectrum and the ability of our eyes to detect the colors.
- *Translucency* – The amount of light reflected and the spectrum of light reflected from the object and detected by the eye, will depend on the ability of the light to travel through the material, where it will change due to absorption and scattering properties of the material and the background against which it is held.
- *Surface texture* – Light can be reflected from a surface, as from a mirror, or scattered in all directions. In the first case the surface is an ideal reflecting polished surface, while in the second case is a matte scattering surface.

The translucency is one of the primary factors in controlling esthetics and a critical consideration in the selection of materials for dental ceramic applications. However, an increase in crystalline content, while achieving greater strength, generally results in greater opacity. If the majority of light passing through a ceramic is intensely scattered and diffusely reflected, the material will appear opaque. If only part of the light is scattered and most is diffusely transmitted, the material will appear translucent. The amount of light absorbed, reflected, and transmitted depends on the amount of crystals within the core matrix, their chemical nature, and the size of the particles compared to the incident light wavelength. The scattering of light in a glass–ceramic is a function of the relative refractive index of the phases.

2.4 Refractive Index

The use of glasses or glass ceramics in dental restorations requires knowledge of the refractive index and how this will affect the translucency of the ceramic. For example in the construction of glass ceramic veneers for metal substructures, the color of the metal needs to be masked by an opaque layer before the more translucent and more esthetic layers are laid down. The higher the opacity, the greater the hiding power and the thinner the layer of opaque glass ceramic that can be used effectively, leaving more room for the more translucent esthetic ceramic layers.

The refraction of light by mineral phases suspended in the clear glassy matrix is the main reason for opacity. The opacity can thus be controlled by a difference in refractive index between two phases in the glass ceramic, which can be as small as 0.05 from the glass. Since all opacifiers (see Table 2.1) have high indices of refraction, it might be concluded that a glass with a low index of refraction would be the most desirable for developing the opacity.

The index of refraction tends to increase with the specific gravity and the degree of opacity depends not only on the relative refractive index of the phases, but also

Table 2.1 Examples of indices of refraction of ceramics

Material	Refractive index
Leucite	1.5058
K-Feldspar	1.52–1.53
Albite	1.535
Quartz	1.55
Lithium disilicate	1.55
Apatite	1.63–1.64
Spinel	1.712–1.736
Aluminum oxide	1.766–1.774
Zirconium oxide	2.15–2.18
Zinc oxide	2.4
Titanium oxide	2.7

on other factors including the particle size of the mineral phases dispersed in the glass and the degree of particle size distribution. So, the smaller the particle size and uniform the dispersion, the more effective the opacity. The opacity is found to increase as the number of reflecting particles per unit volume increases. Also the degree of opacity is affected by the nature of grain boundaries between the opacifying particles and the hosting matrix.

However, the translucency of dental glass ceramic depends on several factors namely the refractive index of the developed mineral phases, their content in the glass ceramics, the refractive index of the residual glassy phase and its content in the glass ceramics.

There are several mineral phases such as leucite and lithium disilicate that have more or less the same refractive index as the glassy phase and thus can be produced with high degrees of translucency. Thus, the crystallization of these mineral phases in the glassy matrix will never have as great an influence on the translucency of the developed glass ceramics as would be the incorporation of some of the metal oxides such as titania.

The leucite crystals have a refractive index that is very close to that of the feldspathic glass. It is this fact that helps to make leucite based dental ceramics esthetically the best ceramics for dental applications.

2.5 Chemical Solubility

In order to be able to elucidate the chemical solubility in dental ceramics, it is important to understand the conditions that directly affect the ceramics. First one has to consider the type of response one seeks to achieve. If the material is to survive for a long time then a low chemical solubility is desirable, whereas if the material is meant to resorb then a high, but controlled solubility is desirable.

Thus for dental ceramics to survive not only do they need to be strong and tough to resist the biting forces, they also have to be able to resist the acidic/alkaline corrosive environment of the oral environment. The ISO standard for dental ceramics states that for ceramics in direct contact with the fluids in the oral environment the ceramic must have a chemical solubility of $<100\ \mu g/cm^2$ after soaking 16 h in 4% boiling acetic acid. So, the only reliable solution is to search for mineral phases to be developed in glass ceramics that provide it with the appropriate characteristics.

The method for measuring the chemical solubility involves the production of ten disk specimens; 12 mm in diameter and 2 mm thick. The samples need to be polished to a smooth surface finish using different grades of abrasives. The samples are washed and dried at $150 \pm 5°C$ for 2 h and weighed and the total surface area is calculated. The specimens are then placed in 4% acetic acid solution (analytical grade) while refluxing for 16 h at 80°C. The samples are rinsed and dried at $150 \pm 5°C/2$ h and reweighed to calculate the chemical solubility, which is presented in terms of micrograms per square centimeter.

In order to show the importance of chemical solubility as a factor controlling the use of a ceramic in the oral cavity, an interesting example is the fluorocanasite glass ceramic. Although the canasite monophase glass ceramic has a high strength and toughness and good microhardness, its Achilles heel is that it has a low chemical solubility. This problem makes canasite unsuitable as a glass ceramic candidate for the construction of dental restorations that come in direct contact with the oral fluids. According to the British standard of dental ceramics, the glass ceramic to be used in the construction of a ceramic core must have chemical solubility of $<2{,}000\ \mu g/cm^2$.

Unfortunately, the chemical solubility of canasite glass ceramics is found to be anything up to $4{,}000\ \mu g/cm^2$. Additionally, the tested samples show an extensive decrease in strength of the samples after soaking in acetic acid and looks like a chalk. The excellent mechanical properties makes it worthwhile to extensive research canasite to improve its chemical solubility to $<2{,}000\ \mu g/cm^2$ for it to be used as a core material and better still $<100\ \mu g/cm^2$ to be able to use it as a single unit restoration without the need to veneer it.

Whereas in the case of a load bearing dental ceramic the chemical solubility wants to be as low as possible, this is not so for bioresorbable ceramics that are used for such biomedical applications as drug delivery, temporary fracture fixation, or for the fabrication of temporary tissue engineering (TE) scaffolds. Depending on the desired function these may remain in the body for weeks, months, or even years. Accurately predicting and evaluating the degradation rate of these materials is critical to their performance, especially if it involves the controlled release of bioactive agents.

One example is the need to use antimicrobial agents that are effective against biofilm-complex aggregations of microorganisms, which are believed to be involved in a wide variety of microbial infections in the body. Phosphate based glasses in the ternary P_2O_5–CaO–Na_2O system are resorbable and this can be controlled by the CaO content. Those incorporating silver ions have potential as temporary antibacterial devices. By controlling the degradation of these glasses, silver ions can be released slowly and continuously over time when in contact with an aqueous medium.

2.6 Mechanical Properties

Glasses and ceramics are very brittle, so it is very important to show the effect of different parameters on the mechanical strength. There are three types of mechanical strengths, which are distinguished according to the way the force is applied. The three types of mechanical strengths include the tensile strength, bending (flexural) strength, and the compressive strength.

On taking tensile strength as unity, the bending strength is approximately twice as high as the tensile strength and the compressive strength is approximately ten times as high as the tensile strength. As the tensile strength is the weakest type of strength, the fracture generally occurs at the point in which the glass ceramic is subjected to a concentrated tensile stress.

In brittle materials, failure takes place at much lower tensile stresses compared with compressive stresses. The strength of a glass and glass ceramic depends to a large degree upon its surface conditions. Tiny cracks in the surface lead to weakening and failure by brittle fracture. Glass and glass ceramics also become weaker with time when subjected to the application of a cyclic stress (Fatigue).

The mechanical properties of glass ceramics are evaluated using a tensile tester and the associated software such as the one shown in Fig. 2.1. Three point bending strength, biaxial flexural strength, and fracture toughness can be measured. The machine deforms a specimen until it breaks, measuring the force required and the deformation at which the break occurred and record stress and strain too if the specimen dimension are measured.

Most applications of bioceramics require them to be used as load bearing structures, especially when used as a bone substitute material or a dental restorative material (see Table 2.1). Thus, the mechanical properties that may be considered of immediate relevance to the clinical use of bioceramics include strength, toughness, and hardness.

When measuring the strength of a ceramic, what can be obtained is not just one value for strength but a range of values, that is, a distribution of strengths, as shown

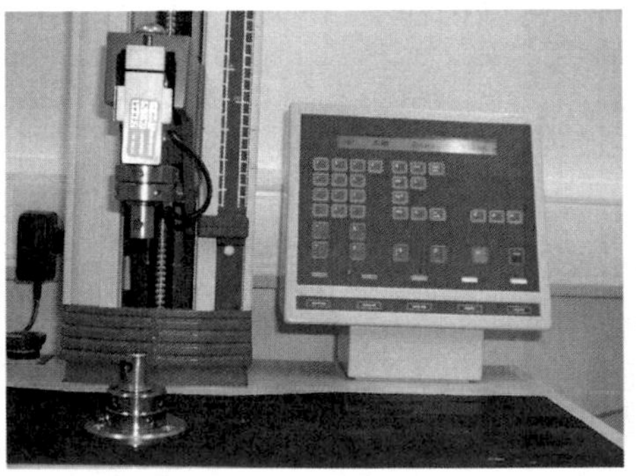

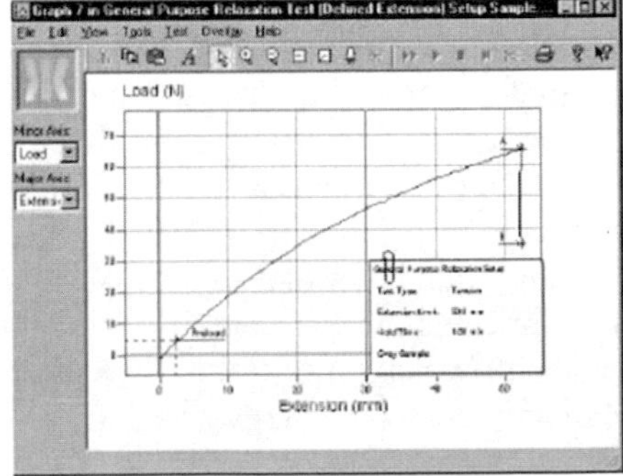

Fig. 2.1 Example of a tensile tester and associated software

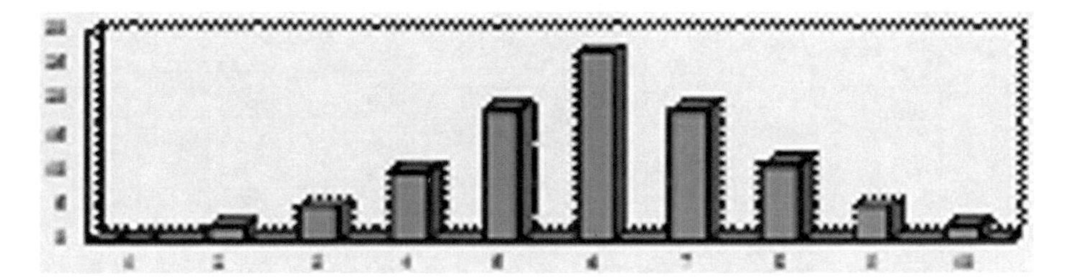

Fig. 2.2 Normal strength distribution curve for a ceramic

in Fig. 2.2, although what is often only reported is the average strength. The reason for this is that the strength is determined by the size, number, and orientation of structural defects. These defects can consist of cracks or holes within the ceramic structure or anything else that can create a local stress concentration. Since each specimen that is tested will have a different arrangement of defects, this will give rise to a different value for the strength.

The other factor that plays an important role is the fracture toughness of the ceramic. Fracture toughness represents the resistance of a material to the propagation of cracks. Thus, when a ceramic contains a crack, the fracture toughness determines how easy it is for this crack to grow. Various mechanisms can be employed to make it more difficult for cracks to grow such as hard or rubbery inclusions or transformation toughening.

Thus it is important to appreciate is that the strength of bioceramics is driven by two factors, namely (1) the inherent flaw size and the (2) the fracture toughness. These three parameters are related by the following equation:

$$\sigma_f = K_{1c}/\sqrt{\Pi a},$$

where σ_f is the tensile strength, K_{1c} is the fracture toughness and a is the inherent flaw size. Hence whenever one seeks to characterize the fracture resistance of a ceramic ideally one should determine both the tensile strength and the fracture toughness.

2.6.1 *Tensile Strength*

There are a variety of methods to determine the tensile strength of a material. These include tensile strength, flexural strength, and biaxial flexural strength. The tensile strength is in many ways the simplest to understand as it is based on the idea that we apply a tensile force to the material until such time that it fails. However, when conducting such an experiment it is important that the experimental design is such that the stress and strain at failure can readily be calculated and for that to be

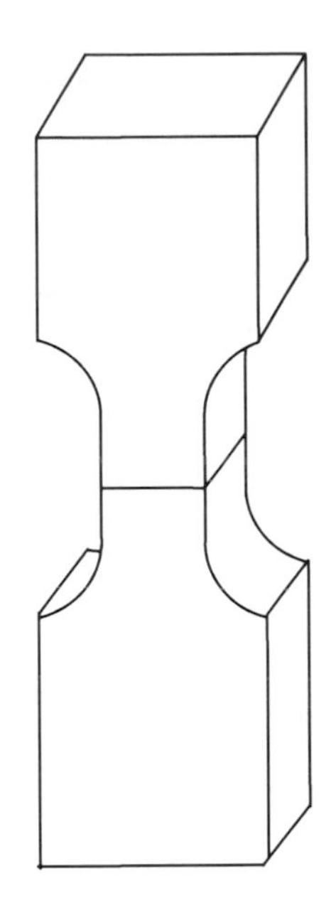

Fig. 2.3 Simple dumbbell design used in tensile strength tests

possible the stress generated within the specimen as a consequence of the applied force needs to be distributed uniformly across the full cross-sectional area of the specimen.

The simplest specimen design that will ensure that is the case is a rod or wire of uniform cross-section. One potential problem with a simple rod or wire is that there is a danger that the specimen could fail at the point at which it is clamped, which would invalidate the result. In order to avoid this problem the specimen design has been modified to provide a thicker cross-section in the clamping part of the testing machine. The design that is most favored is the dumbbell as depicted in Fig. 2.3.

The dumbbell design ensures that the failure is most likely to take place in the central region of the specimen where the cross-sectional area is reduced, such that it is possible to calculate the stress at failure as the force/cross-sectional area (F/A). If the failure were to take place outside this region then the result would have to be discarded. One of the limitations of this design is that for brittle materials, localized high stresses in the clamping zone can cause failure in those areas and consequently many specimens have to be rejected. Thus an alternative and more effective method of assessing the tensile strength of brittle materials needs to be used.

2.6.2 Flexural Strength

The flexural strength test for measuring the tensile strength of brittle materials is based on a simple beam that is subjected to a bending force until such time that it

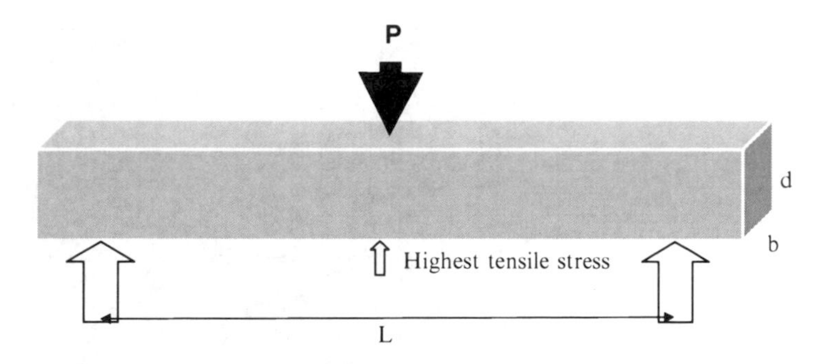

Fig. 2.4 Experimental design for a flexural strength test of a brittle material

fractures (Fig. 2.4). When the beam fails the highest stress will be central to the bottom surface of the beam and is given by:

$$\sigma = 3PL/2bd^2$$

where σ is the stress, P is the applied load, L is the span of the beam between the supports, b is the breadth and d is the depth of the beam. The stress at failure is the flexural strength, also referred to as the modulus of rupture.

Typically, ten standardized bar specimens; 30 mm long and 4 mm width and 4 mm depth are prepared. The specimens are dried and polished to a good surface finish, typically 1 μm, using different grades of abrasives. The specimen is loaded in the central point of the test span at a cross-head speed of 1 mm/min. The load is applied at the midpoint between the two supports by using a knife-edge with a rounded radius of 0.5 mm.

2.6.3 Biaxial Flexural Strength

There are some instances when it is simply easier to make a circular disk of a material with a uniform thickness than it is to make beams, especially as in the case of the beams it is not easy to avoid edge chipping, etc. Many studies of the fracture strength of dental ceramics now prefer to use this method. Essentially the specimen consists of a disk of uniform thickness, typically 12 mm in diameter and approximately 2 mm thick, which is placed on a circular support and loaded centrally with a ball ended loading device (Fig. 2.5).

Typically ten standardized disk specimens, 12 mm long and 2 mm thick, are prepared. The surfaces of the specimens need to be polished very well using different grades of abrasives to produce a consistent surface finish. They can then be centered and loaded on the test machine and tested with a cross-head speed of 1 mm/min. The load is usually applied at the midpoint by a ball shape edge rounded

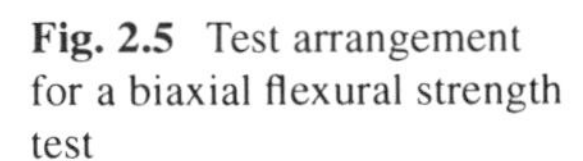
Fig. 2.5 Test arrangement for a biaxial flexural strength test

to a radius of 0.5 mm. When testing is carried out to meet an ISO standard, for the batch to pass the test at least eight of each sample of ten specimens need meet the requirements for strength.

The biaxial flexural strength is determined using the equation:

$$\sigma_{max} = \frac{P^2}{h}\left\{0.606 + \log_e\left(\frac{a}{h}\right) + 1.13\right\}$$

where σ_{max} is the biaxial flexural strength, P is the load to fracture, a is the radius of the knife-edge support and h is the sample thickness.

2.6.4 *Fracture Toughness*

As with strength measurements there are a number of ways in which the fracture toughness of a ceramic can be measured. The methods most commonly used with ceramics are the single-edge notch test and the indentation test.

2.6.4.1 Single-Edge Notch

The fracture toughness can be determined by a single-edge notch three point bend test. This is a variant on the flexural strength test described above in that the distinction is that the surface under tension has been precracked to form a notched bar. The beam will have dimensions that are typically 4 mm wide [*b*], 3 mm high [*d*] and 30 mm long. Using a fine diamond saw, possibly 0.5 mm thick, and with copious

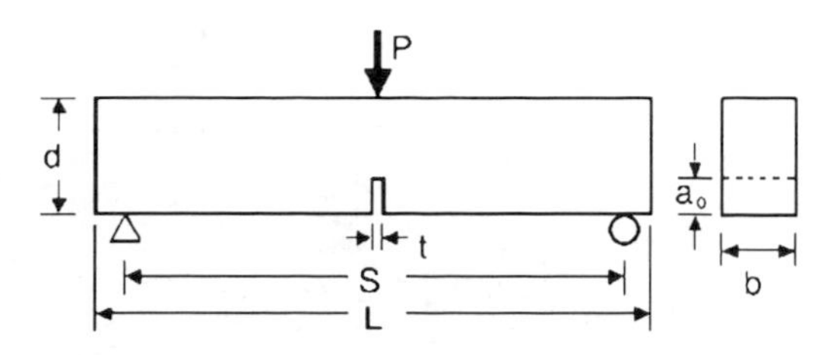

Fig. 2.6 Schematic of Single-Edge Notched beam

water coolant, the specimen is notched in the middle of the 4 mm wide surface to a depth of 1 mm (a_0) (see Fig. 2.6). The notched bars are placed in a three point bend testing arrangement, which will typically have a span length of some 20 mm and are tested at a cross-head speed of 0.5 mm/min. The load at fracture (P) is recorded and the fracture toughness is calculated from the fracture load using the standard equation:

$$K_{1c} = YP/bd^{0.5}$$

where Y is a geometric correction factor.

2.6.4.2 Indentation Method

The fracture toughness can be determined by the indentation technique using a Vickers Hardness tester. The specimens need to have a 1-μm surface finish in order to be able to clearly see the indentation. The specimen is indented under a load that depends on the material and that will ensure that an acceptable crack pattern is obtained. The technique requires that all cracks originate at the corners of the indent and the presence of only four radial cracks with no chipping or crack branching. The radial cracks are measured immediately after each indentation. The fracture toughness is measured using the formula:

$$K_{1c=}\chi P/c^{3/2}$$

where $\chi = 0.0824$, P the indentation load in Newtons, and c the radial crack length in meters.

2.6.5 Microhardness

The hardness is not a precisely defined property. There are a number of methods for hardness determination including the Moh's method (scratch resistance) and the Vickers and Knoop method, which are based on an indentation method and are the ones most commonly used in engineering for precise results.

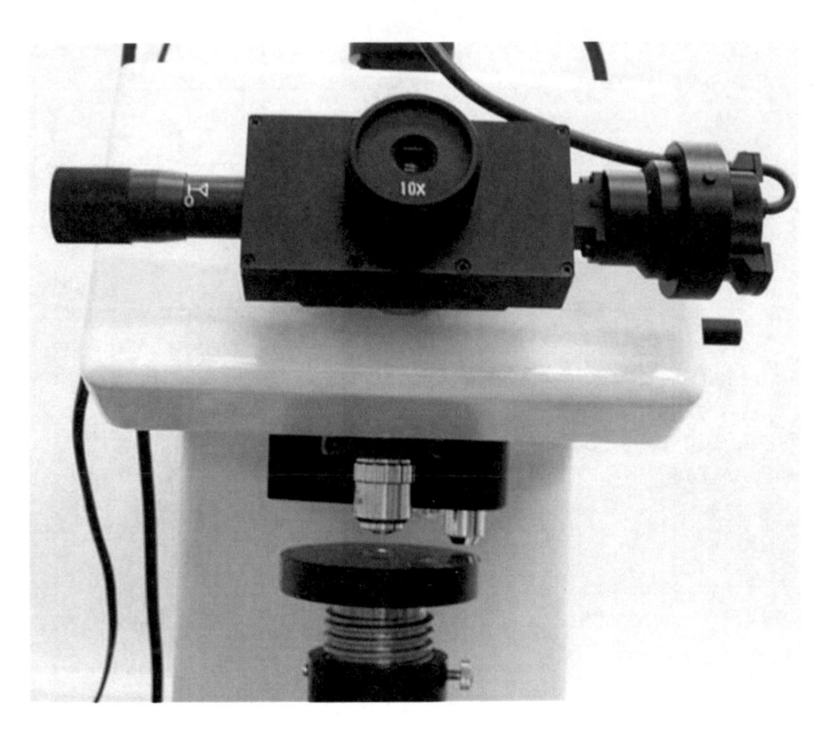

Fig. 2.7 Microhardness indentor

The ability of a harder material to scratch a softer one is the basis for the Moh's scale of hardness, starting from 1 (soft) to 10 (hard). Moh's method assesses the hardness of a material according to which in a series of minerals of successively increasing hardness can scratch the glass and which cannot. The test does not determine the accurate hardness and can only be used for comparison of different glasses.

The most convenient widely used measurement of the hardness of glasses and ceramics is the indentation method. These include the Vickers and Knoop hardness and both use a diamond pyramid indentor (Fig. 2.7).

The Vickers method is advantageous in that the ratio of the hardness numbers corresponds approximately to the actual ratios of hardness of substances. The shape of the indentation is created with a square diamond pyramid. The hardness is determined by measuring the diagonals of the indentation in the glass surface.

For example a substance with a hardness of 400 is four times as hard as a substance with a hardness of 100 HV. The microhardness is calculated from the following equation

$$\mathrm{HV} = 1.854P/d^2 = \mathrm{Nm}^{-2}$$

where P is the force acting on the Indentor in Newtons and d is the diagonal of the rhomb impression in micrometers.

The Knoop hardness is a modification of the Vickers method and differs only in the shape of the indentor in that the shape of the indentation is rhomboid. The Knoop hardness is calculated using the following equation:

$$H_k = 1.4513P/d^2$$

where P is the force acting on the Indentor in Newtons and d is the length of the longer diagonal. The smaller the area of indentation of a fixed load, the harder is the material.

The microhardness depends on the surface finish of the specimen to be measured and until now there is no accurate method for calculating the microhardness of the materials. For this reason, the most useful hardness measurements are made by comparing surfaces treated in the same manner with the same measuring method and load. The indentation method is also used in calculating the fracture toughness.

2.6.6 Machinability

The advent of advanced manufacturing technologies such as CAD-CAM means that there is a growing interest in the machinability of medical ceramics. Particularly in dentistry this has become a major means of producing dental restorations such as veneers, inlays, crowns, and bridges. Some ceramics can be machined directly in its final state such as the feldspathic glasses, whereas other ceramics need a two stage process because the final product is too difficult to machine. For example, partially stabilized zirconia would be very difficult to machine in its final state and therefore a process known as "soft machining" has evolved. This involves producing a porous block of the ceramic, which is easy to machine but can shrink up to 20% on densification.

Once the component has been machined, which takes account of the amount of shrinkage, it is then put through a densification firing cycle. Another example is the machining of components from a glass ceramic such as lithium disilicate. Again in its final state it is difficult to machine and in order to overcome this problem the glass–ceramic is partially crystallized, the component is machined from the partially crystallized block and the final product is then given a further crystallization firing cycle. The advantage is that the second firing cycle does not involve any shrinkage and thus the size of the component is not compromised.

The machinability of glass ceramics is found to be affected mainly by the type of mineral phases in the glass. For example, simultaneous crystallization of fluorophlogopite and beta-spodumene solid solutions results in a machinable ceramic. However, the beta-spodumene phase tends to harden the body by increasing the microhardness and thereby impair the machinability character. The ceramic will be no longer machinable when the beta-spodumene proportion approaches more than 50% (Table 2.2). On the other hand, the beta-spodumene phase will improve the

Table 2.2 Microhardness of beta-spodumene–fluorophlogopite

	Microhardness values (kg/mm²)		
Temp. (°C)	F_1 (10% spodumene)	F_3 (30% spodumene)	F_6 (60% spodumene)
850	50.70	53.90	101.00
900	62.00	96.60	110.20
950	113.50	118.10	117.50
1,000	75.60	168.80	217.00
1,050	80.40	132.90	689.70
1,100	95.60	105.50	911.00

microhardness without impairing the machinability if it occurs in an amount <30%. The two mixes F_1 and F_3 have been shown to be readily machinable glass ceramics. In addition, the presence of beta-spodumene solid solution enables lower expansion coefficient to demonstrate excellent resistance to thermal shock.

The best machinability can be secured at the highest concentrations of fluormica and wherein the fluorophlogopite has plate-like grains with a grain size of less than 5 μm. Mica grains with a uniform size and randomly distributed in the glassy matrix are known to produce glass ceramics with a very good machinability.

The machinability of a ceramic is not easy to quantify. Various parameters have been suggested as a "measurement" of the machinability, such as tool wear, surface roughness, cutting force, cutting energy, drilling rates. A more practical suggestion is that it can be assessed from a calculation of the brittleness index. The brittleness index, as suggested by Lawn and Marshall in 1979, is a ratio of the Vickers hardness to the fracture toughness and is calculated using the following formula:

$$B = H_V / K_{1c}.$$

Boccaccini (1996) has shown that the brittleness of glasses and glass–ceramics can be calculated using the following equation:

$$B = 1.599 P^{-1/4} (c/a)^{3/2}.$$

In this equation, B is the brittleness index in $\mu m^{1/2}$, P the indentation load (N), which should be set at 49 N for median cracking, c is the median crack length and a is the contact diagonal of the indent in mm. A brittleness index lower than 4.3 $\mu m^{1/2}$ is considered to be within the range for good machinability.

2.7 Thermal Behavior

The thermal behavior of glasses and glass ceramics is very important for various applications. Many features result from heating the glass or glass ceramics which are either desirable or undesirable according to the application of the material.

Measuring the thermal behavior of the ceramic is very important during the annealing of glass, coating substrates with glass, or glass ceramics, the crystallization and nucleation of the glass, the thermal shock resistance of the materials, etc. The thermal behavior can be evaluated using both a dilatometer and differential thermal analysis (DTA).

2.7.1 *Thermal Expansion*

When a glass ceramic material is heated, the extra energy absorbed causes the atoms to vibrate with increased amplitude and the ceramic expands. This change in length, when determined per unit length for a 1 °C change in temperature, is called the thermal expansion coefficient (TEC) and is expressed in terms of parts per million per degree centigrade (ppm/°C).

Internal stresses generally generate due to strains resulting from irregular thermal expansion of the ceramic. For example, if a glass ceramic is heated on one face and cooled on the other face, compressive stresses are set up on the hotter face and tensile stresses are generated on the cooler face. Whether or not a uniform temperature gradient is established across the thickness of the glass ceramic depends on the thickness of the sample.

With sudden heating and cooling, very high stresses can be generated resulting in the development of compressive stresses at the surface. Rapid cooling results in a greater initial contraction of the surface layer than the interior of the glass which places the surface under tension. Glass and glass ceramic objects are much more likely to fail if the object is already in a state of tension due to differential thermal contraction.

There are a number of instances where the thermal expansion of the ceramic is a matter of concern. For example, when using a glass ceramic such as a feldspathic glass ceramic containing leucite as a veneering material for a metal substrate, as in dental crowns and bridges, it is vital that the mismatch in the TEC is not too great, otherwise on cooling the ceramic layer may crack or peel and separate from the metallic substrate.

The thermal expansion of a glass ceramic is very important for the thermal compatibility in different applications. If a glass ceramic is needed with a good thermal shock resistance then a low thermal expansion glass ceramic is desirable. On the other hand, a glass ceramic with a high TEC, probably between 13 and 15×10^{-6}/°C, would be needed if it is to be used to conceal a Ni–Cr alloy by acting a surface coating. Thus, by adjusting the TEC this can help to fix various industrial problems.

Another area of increasing interest is the coating of metallic implants with bioactive glasses in order to improve the ability of the inactive implant to bond to living tissue, particularly bone. The TECs of bioactive glasses are typically much larger than those of Ti alloys. Typically the TEC of a bioactive glass is in the region of 15×10^{-6}/°C whereas that for a Ti_6Al_4V alloy it is around 10×10^{-6}/°C. Whilst the

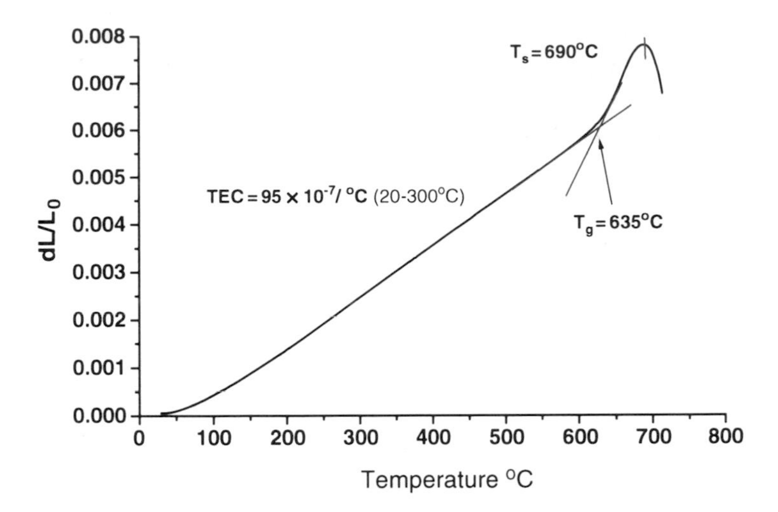

Fig. 2.8 Thermal expansion curve for a leucite glass ceramic

simplest way to reduce the thermal expansion of a bioactive glass is to increase the SiO_2 content and decrease the alkali oxides contents of the glass, this is unfortunately at the expense of the bioactivity, which is significantly reduced.

The thermal shock resistance is a practical example of the importance of expansion measurement which depends mainly on the value of the TEC. The extent to which the material can withstand sudden temperature changes without fracture is referred to as the thermal shock resistance. The thermal shock resistance is defined in terms of the maximum temperature interval through which the material can be rapidly heated and cooled without fracturing. So the thermal shock resistance is important for ceramics exposed to sudden cooling and heating applications.

The expansion curve of a glass is a very important factor in the construction of a crystallization heating schedule for a glass into glass ceramics. Several significant points in the expansion curve of glass can be predicted from the thermal expansion curve of glass, including the lower annealing point, transformation temperature, the upper annealing point, and the softening point in addition to the nature of the melting. The information on T_g and T_s are very important in determining the best crystallization schedule (Fig. 2.8). Also the annealing temperature of the glass during preparation of glass blocks for production of dental glass ceramics for a CAD–CAM machine can be predicted.

The measurement of this expansion is carried out by taking a length of material, heating it to a certain temperature and then measuring the resultant change in length using a dilatometer; Specimens are prepared as rods having a length of 25 mm and cross-sectional diameter not greater than 5 mm. The ends of the specimens are ground to yield flat surfaces to the perpendicular the long axis of the specimen. The samples must be well annealed to avoid the false results.

The specimen is heated at a heating rate of 5°C/min between room temperature and the softening point to determine the TEC, transition temperature, and the

Fig. 2.9 DTA equipment

softening temperature. The glass transition temperature is estimated graphically for each ceramic specimen from the plotted curves of expansion versus temperature. The samples are prepared using the above mentioned preparation method and fired using the above firing schedule.

2.7.2 Differential Thermal Analysis

The chemical reactions or structural changes within the amorphous glass and the crystalline glass ceramics are accompanied by the evolution or absorption of energy in the form of heat. For example when a glass substance crystallizes, an exothermic effect occurs since the free energy of the developed regular crystal phase is less than that of the structurally disordered glass substance. The melting of crystalline glass ceramics is represented by an endothermic reaction as the liquid state needs a higher free energy than that of the glass.

Also it can be predicted if the glass ceramic contains a monocrystalline phase or multiphase crystalline materials from the shape of endothermic peak. Thus DTA is a technique that facilitates the study of phase development, decomposition, or phase transformations during the glass ceramic process (Fig. 2.9).

In this method the glass material under test is in the form of a finely divided powder placed in a small crucible of platinum or other suitable refractory crucible made of aluminum oxide. Adjacent to the test crucible is a second crucible containing an inert powder such as aluminum oxide, which does not exhibit endothermic or exothermic reactions.

The temperature difference between the tested sample and the reference sample powder (almost always inert alumina) is recorded during heating. The DTA curve

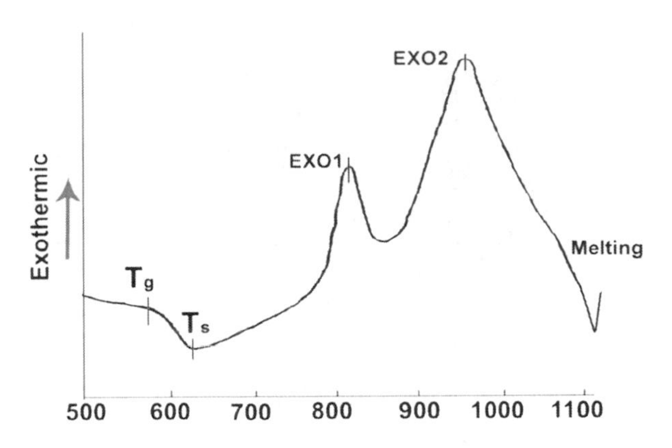

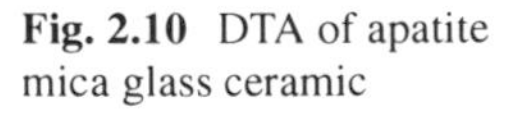
Fig. 2.10 DTA of apatite mica glass ceramic

is plotted. The exothermic reactions are represented as peaks and the endothermic reactions are represented by dips on the DTA curve.

The technique is considered a useful method for the characterization of the glasses and the determination of the optimum temperatures at which the different crystal phases are formed. The quality and the shape of the peaks are affected by the rate of heating and the fineness of the powder.

As the temperature is increased, a dip is observed on the DTA curve due to slight absorption of heat that occurs when the annealing point of the glass is reached. With further increase of the temperature, one or more quite sharp exthothermic peaks are expected, corresponding to the development of various crystal phases. At a very high temperature the endothermic reaction peak relates to the early melting of the tested crystalline sample. The main features of a typical DTA curve of a glass are shown in Fig. 2.10. The curve shows the thermal behavior of a glass that can crystallize to yield two phase glass ceramics.

Therefore the DTA analysis yields a great deal of very useful information which is of great help in designing the heat treatment schedules for the crystallization of a glass into glass ceramics. The curve not only indicates the temperature range in which the nucleation and crystallization occurs, but also it indicates the maximum temperature to which the glass can be heated to prepare glass ceramics without exposing the glass ceramic to deformation due to melting of the crystal phases. Having determined the exothermic and endothermic peaks, the crystallization of the different crystalline phases can be perfectly assessed. In cooperation with the XRD data, one is able to assess the best schedule for a crystallization of a certain phase or phases.

Chapter 3
Grouping of Ions in Ceramic Solids

In this chapter, the nature of the atomic arrangements, the forces between atoms, and the location of atoms in the crystalline ceramics and noncrystalline glass are considered. As the silicates form the main chemical constituents of the glasses and glass ceramics, and porcelains discussed in this book, the structure of silicates is explained briefly to clarify the grouping of different ions in either the glasses or the ceramic. In addition, the rules governing the coordination of cations and anions are briefly discussed.

3.1 Ceramic Solids

Ceramic solids are single crystal, polycrystalline, wholly amorphous, or mixtures of crystalline and glassy phases. Pore spaces are also considered an important constituting phase in most ceramic materials. Some applications need porous ceramics with controlled pore size distribution as in porous bioactive glass ceramics or porous hydroxyapatite for substituting bone with sufficient strength for medical implants.

Both the crystalline and amorphous solids differ in their structures, where almost all ceramic materials vary from perfectly crystalline to amorphous with random arrangement. The range includes all types of ceramics as well as glass ceramics. This fact will be tracked as a base for understanding of the properties of crystalline and noncrystalline solids.

Almost all atoms of the elements in the periodic table can be employed in building the structure of ceramics. However, as we are concerned only with medical ceramics, the range of elements under investigation should be physiologically nontoxic and nonradioactive, when incorporated in the structure of ceramics.

E. El-Meliegy and R. van Noort, *Glasses and Glass Ceramics for Medical Applications*,
DOI 10.1007/978-1-4614-1228-1_3, © Springer Science+Business Media, LLC 2012

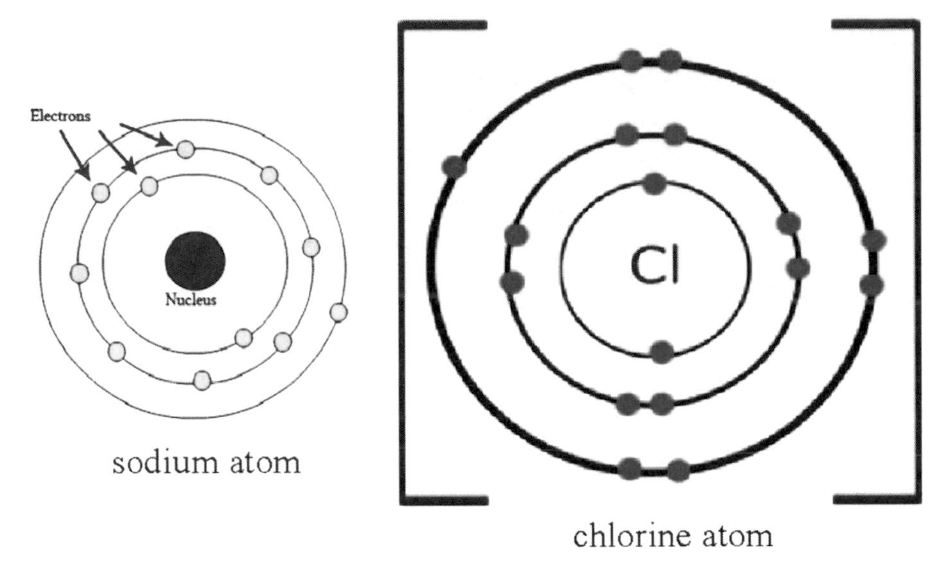

Fig. 3.1 The atomic structures of the sodium and chlorine atoms

3.2 The Structure of the Atom

The atom is the smallest unit of a substance that defines its chemical properties. All atoms have the same basic structure: a nucleus surrounded by negatively charged electrons orbiting around the nucleus. The nucleus contains both positively charged protons and an equal number of neutral neutrons. Protons are similar in masses to neutrons. Each atom corresponds to a unique chemical element, which is determined by the number of protons in its nucleus. The atomic number of an element is equal to the number of electrons orbiting around the atom or the number of protons in the nucleus. The structures of the sodium and chlorine atoms are shown in Fig. 3.1.

3.3 Formation of Ions and Ionic Compounds

Atoms are known to be neutral because each atom contains the same number of protons and electrons. The ion is formed either by removing or gaining one or more electrons. When a neutral atom loses electrons, it becomes a positively charged ion (cation). Alternatively, when the neutral atom gains electrons, it becomes a negatively charged ion (anion). The number of protons does not change when the atoms lose or gain electrons to become an ion.

A neutral sodium atom, for example, contains 11 protons and 11 electrons. By removing an electron from the sodium atom, it becomes a positively charged Na^+ ion. Atoms that gain one or more of electrons become negatively charged ions.

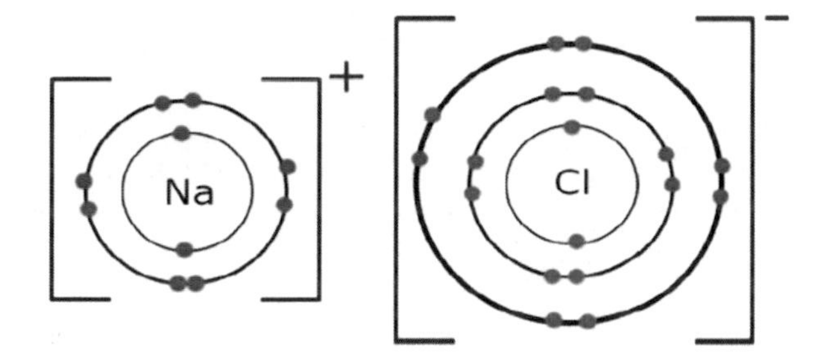

Fig. 3.2 Sodium chloride ionic bond, Kingery et al. (1976)

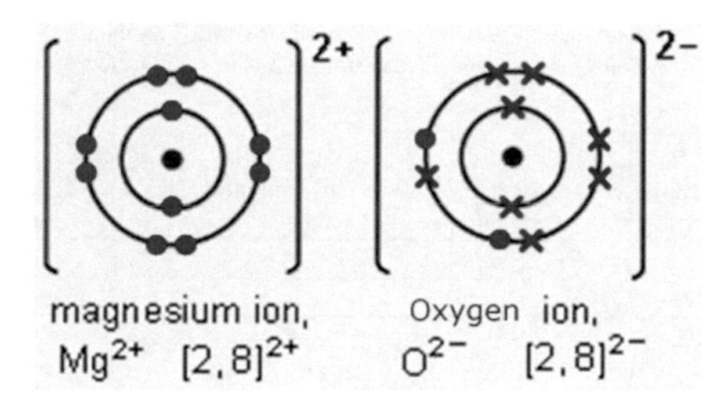

Fig. 3.3 MgO ionic bond formation

Thus, a neutral chlorine atom contains 17 protons and 17 electrons. By adding one more electron, a negatively charged Cl^- ion is formed.

When a sodium atom loses its outer shell electron and become a positively charged Na^+ ion, the chlorine atom captures the freed electron to complete its outer shell with eight electrons (Fig. 3.2). The process results in an ionic bond, which is formed between the sodium atom and chlorine atom resulting in a sodium chloride compound (NaCl).

A neutral magnesium atom contains 12 protons in its nucleus and 12 electrons orbiting around the nucleus. By removing two electrons from the outer shell of the magnesium atom, the magnesium atom becomes positively charged Mg^{2+} ion. On the other hand, neutral oxygen atom, for example, contains eight protons and eight electrons. By gaining the two freed electrons released from the magnesium atom, a negatively charged O^{2-} ion is formed as shown in Fig. 3.3.

3.4 The Ionic Size

The arrangement of ions in a crystal structure depends not only on the ionic charge but also on the size of ions. The size of ions depends on the size of nuclei and the atomic number. Atoms with higher numbers of electrons have larger radii than those with smaller number of electrons. So, the radius of an ion may change on losing or gaining electrons. An ion that becomes positively charged, will be smaller in radius as it loses electrons, while the ion that gains electrons becomes larger.

3.5 Coordination Number

The coordination number (CN) of a cation is represented by the number of anions surrounding the cation in the structure. The ionic radii of the most common elements used in ceramics and their changes with changing the coordination numbers are shown in Table 3.1.

The ionic radii are found to increase with increasing coordination number and vice versa. A degree of generalization for the arrangement of ions in the crystalline structure has been expressed by Pauling's rules. The first rule states that:

> a coordination polyhedron of anions is formed around each cation in the structure and the cation–anion distance is determined by the sum of their ionic radii.

Table 3.1 The ionic radii of the most common elements used in medical glass ceramics and their changes with changing the coordination structure

	Radius (Å)				
Ions	CN=4	CN=5	CN=6	CN=7	CN=8
Al^{3+}	0.39	0.48	0.54		
Ag+	1.00	1.09	1.15		1.28
B^{3+}	0.11		0.27		
Ba^{2+}			1.35	1.36	1.42
Ca^{2+}			1.00	1.06	1.12
Ce^{3+}			1.01	1.07	1.14
Co^{2+}	0.58	0.67	0.75		0.90
Er^{3+}			0.89	0.95	1.00
F^{-}	1.31		1.33		
K^{+}	1.37		1.38	1.46	1.51
La^{3+}			1.03	1.10	1.16
Li^{+}	0.59		0.76		0.92
Mg^{2+}	0.57	0.66	0.72		0.89
Mn^{2+}	0.66	0.75	0.67	0.9	0.96
Na^{+}	0.99	1.00	1.02	1.12	1.18
O^{2-}	1.38		1.4		1.42
P^{5+}	0.17	0.29	0.38		
Si^{4+}	0.26		0.4		
Sn^{4+}		0.62	0.69	0.75	0.8
Ti^{4+}	0.42	0.51	0.61		0.74
Tb^{3+}			0.92	0.98	1.04
Sb^{3+}	0.74	0.80			
V^{5+}	0.36	0.46	0.54		
Y^{3+}	0.90			0.96	1.02
Yb^{3+}			0.87	0.93	0.99
Zn^{2+}	0.6	0.68	0.74		0.90
Zr^{4+}	0.59	0.66	0.72	0.78	0.84

3.6 Electronegativity

Electronegativity is the ability of an atom to attract electrons from another atom in an attempt to fill its outer electronic shell. The higher the electronegativity of an atom, the closer it pulls the electrons, the more strongly an atom attracts electrons. So the ionic size is also affected by the electronegativity.

3.7 Bonding of Ions in Ceramic Solids

Bonding in ceramics is mostly ionic. The charge neutrality is shown to be zero for stable compounds. The forces between ions in ceramic solids result in the coordination structure of anions around the cations to form the periodic array of a unit cell. The unit cell is the building unit of the crystal. The sodium chloride unit cell is shown in Fig. 3.4. Various unit cells fit together to form the crystalline structure.

An understanding of the properties of crystalline ceramic is made easier if we know the fundamentals of how the periodicity in crystals is obtained and the possible substitution of different ions in the structure and the rules controlling ionic substitution. Stable crystal structures are found to have the densest packing of atoms that is consistent with ionic sizes and bonding.

The grouping of ions in crystals having a large structure, as in the case of silicates, is determined on the basis of how positive and negative ions can be packed to maximize the electrostatic attractive forces and minimize electrostatic

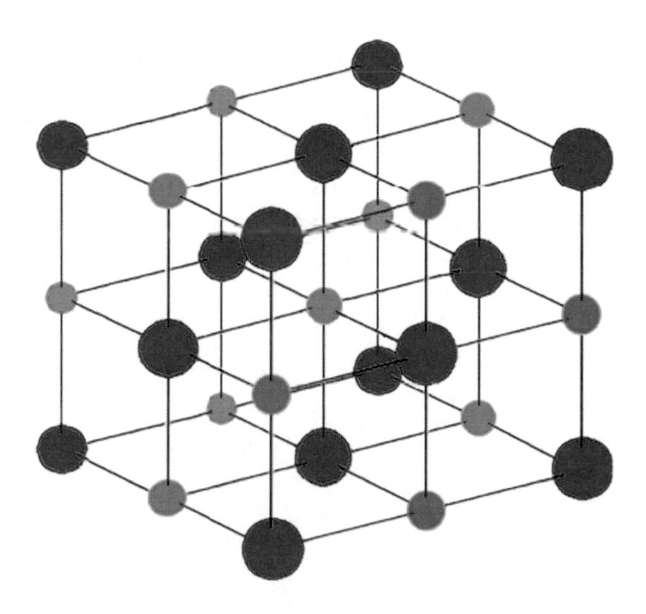

Fig. 3.4 Sodium chloride unit cell

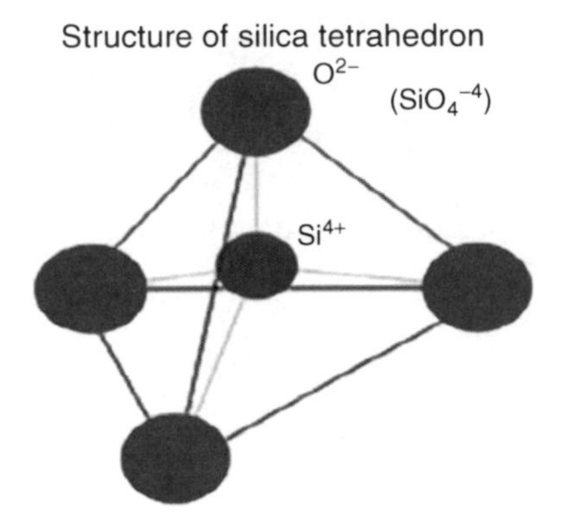

Fig. 3.5 A silica tetrahedron

repulsive forces. The electrostatic repulsive forces decrease the more the ceramic deviates from randomness to periodicity to allow bonds to exist. The stable array of ions in a crystal structure is the one that has the lowest energy.

3.8 The Ionic Bond Strength

Pauling's second rule describes a basis for evaluating the electrical neutrality of the local structure. It states that "in a stable crystal structure, the total strength of bonds reaching an anion from all of the surrounding cations should be equal to the charge of the anion."

The strength of an ionic bond given from a cation to anion is defined as the formal charge of the cation divided by its coordination number. For example, silicon with valence 4 forms a tetrahedral coordination and has bond strength $4/4=1$ as shown in Fig. 3.5. Also aluminum with a valence of 3 has octahedral coordination with a bond strength of $3/6=1/2$.

For example, in the $Si_2O_7^{6-}$ unit, two bonds of strength 1 reach the shared oxygen ion from the surrounding silicon ions, thus the sum of the bonds is 2, which represents the valence of the oxygen ion. Similarly, the structure of spinel ($MgAl_2O_4$) contains one Mg^{2+} cation that has a bond strength of 2/4 and three Al^{3+} that have three bonds, each of them has a bond strength of 3/6.

3.9 Prediction of the Ionic Packing Structure

To predict the coordination polyhedron, we need to examine the relative sizes of the ions to see what will happen when one of the involved ions becomes smaller in size or larger in size. The relative sizes are indicated by the radius ratio of the coordinating ions. In the crystal structures the cations are surrounded by anions and the radius ratio is defined as the ratio of the radius of the cation to the radius of the anion R_{cation}/R_{anion}. The coordination number depends on the relative sizes of the ions. Let us imagine that the ions in a crystal are of the same size, in this case, the maximum number of anions that is coordinated around any individual cation is 12. So this 12-fold coordination forms a dodecahedron packing structure.

Since the anions are usually larger in size than cations, this results in reducing the value of the radius ratio. With decreasing size of the cations, the coordination number will be lower and then different cation–anion coordination packing is possible. So we can predict what coordination number and coordination structure can be formed in a glass ceramics through calculating the radius ratio of the ions involved in the structure. This fact is very important for understanding ceramics as almost all ceramic phases are based on cation–anion coordination, cation–cation substitution, and anion–anion substitution in the crystalline structure.

The coordination number is determined by the radius ratio. A given coordination number is thus stable only when the radius ratio is greater than some critical value. In a crystal structure, the anion is also surrounded by a coordination of cations. Critical radius ratios also govern the coordination of cations about the anions. The critical radius ratio for a structure is almost always determined by the coordination of the anions around the cations. Since the radii of anions are generally larger than the radii of cations, the radius ratio is generally less than one. The limits of critical values for the radius ratio are given in Table 3.2.

Oxygen is the major anion that coordinates the other cations in ceramic solids. Table 3.3 lists the different coordination structures of different cations calculated based on the radius ratio and shows the expected coordination and coordination polyhedra.

Table 3.2 Different coordination structures and the corresponding critical radius ratios

Radius ratio	Coordination number	Coordination structure	Examples
0–0.115	2	Linear	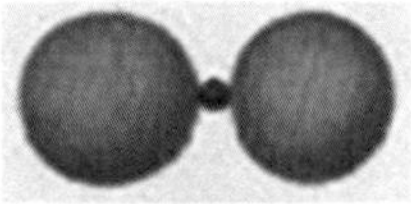
0.155–0.225	3	Triangular	
0.225–0.414	4	Tetrahedral	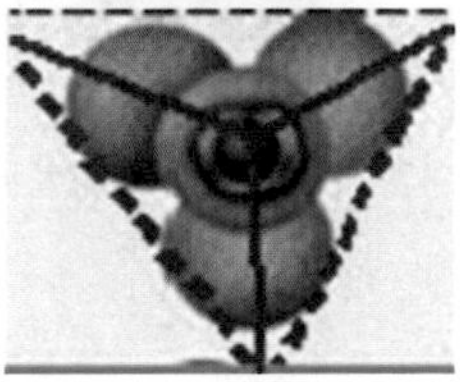

(continued)

Table 3.2 (continued)

Radius ratio	Coordination number	Coordination structure	Examples
0.414–0.732	6	Octahedral	
0.732–1.00	8	Cubic packing	
1.00	12	Hexagonal to cubic	

Table 3.3 The coordination structures of different cations calculated based on the radius ratio with the radius of oxygen being 1.4 Å

Ions	Radius of cations, Å	Radius ratio	Coordination number	Coordination polyhedra
Al^{3+}	0.54	0.39	6	Octahedron
Al^{3+}	0.39	0.28	4	Tetrahedron
B^{3+}	0.11	0.19	3	Triangle
Ca^{2+}	1.12	0.80	8	Cubic
Ca^{2+}	1.00	0.71	6	Octahedron
Ce^{3+}	1.01	0.72	6	Octahedron
Co^{2+}	0.58	0.41	4	Tetrahedron
Er^{3+}	0.89	0.64	6	Octahedron
Fe^{3+}	0.65	0.46	6	Octahedron
Fe^{2+}	0.78	0.56	6	Octahedron
K^{+}	1.38	0.99	8	Cubic
Mn^{2+}	0.83	0.59	6	Octahedron
Mg^{2+}	0.72	0.51	6	Octahedron
Na^{+}	1.18	0.84	8	Cubic
Na^{+}	1.02	0.73	6	Octahedron
Si^{4+}	0.4	0.29	4	Tetrahedron
Sb^{3+}	0.74	0.53	6	Octahedron
Ti^{4+}	0.61	0.44	6	Octahedron
V^{5+}	0.36	0.26	4	Tetrahedron
Y^{3+}	0.90	0.64	6	Octahedron
Yb^{3+}	0.87	0.63	6	Octahedron
Zn^{2+}	0.6	0.43	6	Octahedron
Zr^{4+}	0.59	0.42	6	Octahedron

3.10 Stability of the Coordination Structure

The stable structure always has the maximum permissible coordination number, where the electrostatic energy of a crystal is obviously decreased. So the most stable structure is usually the one with the maximum coordination number allowed by the corresponding critical radius ratio. Some experimentally observed coordination numbers are compared with predicted values for various cations with oxygen anions in Table 3.4.

Pauling's third rule is further concerned with the geometrical linkage of the cation coordination polyhedra and says that:

> in a stable structure, the corners, rather than the edges and the faces of the coordination polyhedra tend to be shared.

As the separation of the cations within the polyhedron decreases, the polyhedra successively share corners rather than edges and faces. In this case, the repulsive interaction forces between cations accordingly increases. Pauling's fourth rule states that:

> Polyhedra formed about cations of low coordination number and high charge tend to be linked by corner sharing.

The repulsive interaction between a pair of cations increases and the separation of cations within a coordination polyhedron decreases as the coordination number of the cation becomes smaller. Pauling's fifth rule states that:

> the number of different constituents in a structure tends always to be small to reach better stability.

This arises from the difficulty encountered to reach efficient packing into a single structure.

Table 3.4 Comparison of some experimental and predicted coordination numbers

Ions	Radius (CN=6)	Radius ratio	Predicted CN	Observed CN	Bond strength
B^{3+}	0.23	0.16	3	3, 4	1 or ¾
Li^{+}	0.74	0.53	6	4	¼
Si^{4+}	0.40	0.29	4	4, 6	1
Al^{3+}	0.53	0.38	4	4, 5, 6	¾ or ½
Mg^{2+}	0.72	0.51	6	6	1/3
Na^{+}	1.02	0.73	6	4, 6, 8	1/6
Ti^{4+}	0.61	0.44	6	6	2/3
Ca^{2+}	1.00	0.71	6, 8	6, 7, 8, 9	¼
Ce^{4+}	0.80	0.57	6	8	½
K^{+}	1.38	0.99	8, 12	6, 7, 8, 9, 10, 12	1/9

3.11 Solid Solutions

Glass ceramics are formed from one or more crystalline phases dispersed in the glassy matrix. The mineral phase has a definite but not necessarily fixed chemical composition. Elements substitute for each other in the crystalline structure according to their ionic size, radius ratio, and ionic charge. The variation in the chemical composition due to replacement of a cation in the crystalline structure is called solid solution. So there exists a solid solution similar to what happens in the liquids. We can also describe solid solution as dissolution of one solid mineral phase in another phase. Solid solution extent is affected by several factors that control its amount:

1. The sizes of ions substituting each other; generally ions with the same size can substitute for each other.
2. The size of the crystallographic sites into which they substitute since the size of the crystallographic site can also play a role if one of the ions is nearly of the same size, but is too large to fit into the crystallographic vacancy.
3. The ionic charges of both elements; cations with the same charge substitute for each other easily and maintain the electrical neutrality of the crystal structure, while cations with different charges will require other substitutions in order to maintain the charge balance.
4. The effect of temperature and/or pressure can result in occurrence of other substitutions irrespective of realizing all of the above conditions, where both the size of ion and the crystallographic site are affected by the temperature and pressure.

There are different types of solid solution, including substitutional, coupled, and interstitial solid solutions.

1. Substitutional solid solution
   In a substitutional solid solution (Fig. 3.6) ions of nearly equal sizes and equal charge simply substitute each other. This solid solution is either complete or

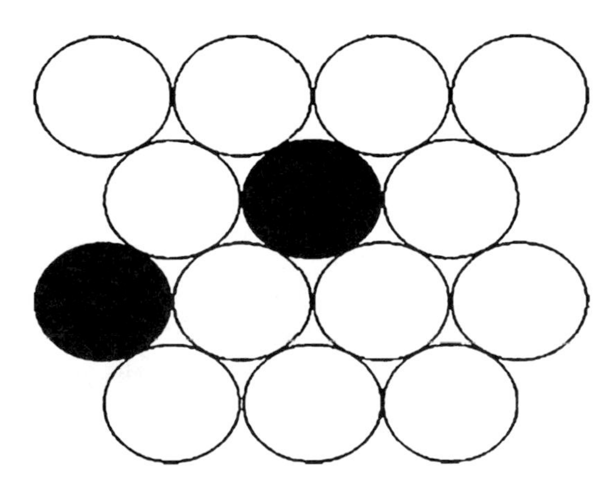

Fig. 3.6 Substitutional solid solution

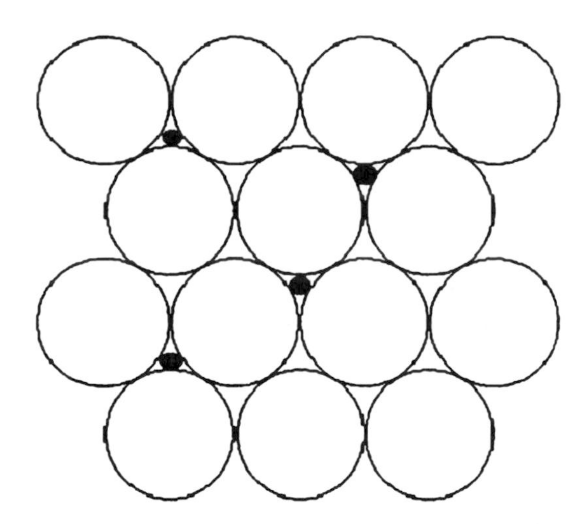

Fig. 3.7 Interstitial solid solution

partial. Complete solid solution means that the two cations can completely substitute each other in the crystalline structure. Partial solid solution means that the two cations can substitute each other in the structural to certain limit.

Albite ($NaAlSi_3O_8$) and orthoclase ($KAlSi_3O_8$) form a solid solution, where Na^+ with an ionic size of 1.18 Å and K^+ with an ionic size of 1.51 Å can substitute each other partially at low temperature as both of the cations have the same charge but are slightly different in ionic size. This size difference is overcome at higher temperatures where a solid solution across the full range of compositions can exist.

2. Coupled substitutional solid solution
When ions of different charges are substituted, this requires another substitution in order to maintain the charge balance. This coupled substitution is common in the silicate structures, such as, for example, Al^{3+} substituting for Si^{4+}, which is followed by a Ca^{2+} substituting for a Na^+ ion to maintain the charge balance. Also, in the case of albite ($NaAlSi_3O_8$) and Anorthite ($CaAl_2Si_2O_8$), a complete solid solution is possible following the formula $(Ca_{1-x}\ Na_x)(Al_{2-x}Si_{2+x})O_8$, when the $Na^+\ Si^{4+}$ couple can completely substitute Ca^+Al^{3+} couple and similarly a $Ca^{2+}Mg^{2+}$ couple can substitute Na^+Al^{3+} in the case of diopside ($CaMgSi_2O_6$) and Jadeite ($NaAlSi_2O_6$).
3. Interstitial solid solution
In an interstitial solid solution (Fig. 3.7), ions do not displace atoms in the crystalline structure, but rather occupy the interstices between the atoms in the crystalline structure to occupy an interstitial position between iron atoms. Normally, atoms, which have atomic radii less than 1 Å, are likely to form interstitial solid solutions.

Creating interstitial cations: when the substituting ion has a lower charge than the hosting ion, an interstitial cation is required to maintain charge balance. When Al^{3+} substitutes Si^{4+} in SiO_2, a Li^+ interstitial is required, $Li_x(Si_{1-x}\ Al_x)O_2$.

3.12 Model of Solid Solutions

The substitution of elements in feldspathic glasses represents an excellent case study for showing the complete solid solution phenomenon. The compositions of the major elements in common feldspars can be expressed in terms of three end members:

- K-feldspar $KAlSi_3O_8$
- Albite $NaAlSi_3O_8$
- Anorthite $CaAl_2Si_2O_8$

Two are alkali feldspars in which the potassium cation replaces the sodium cation in the structure of the albite mineral to form K-feldspar and the third in the series is a plagioclase feldspar (anorthite) in which the calcium cation replaces the sodium cation in the structure of albite phase. All of them are based on the silica tetrahedra, which form rigid three-dimensional interlocking frameworks (tectosilicates) in feldspars. Anorthite has only a limited solid solution in K-feldspar and albite. In addition, barium feldspars can form as potassium is replaced by barium in K-feldspar. The barium feldspars are monoclinic and comprise either celsian ($BaAlSi_3O_8$) or hyalophane $(K,Na,Ba)(Al,Si)_4O_8$.

The plagioclase feldspars are triclinic and the composition can range from the albite mineral phase $Na[AlSi_3O_8]$ to the anorthite mineral phase $CaAl_2Si_2O_8$ and any mixture in between as Ca^{2+} will substitute for Na^+ in albite at higher temperature The plagioclase series are shown in Table 3.5.

The potassium feldspars can exist as orthoclase (monoclinic) or microcline (triclinic). Microcline forms during slow cooling of orthoclase; it is more stable at lower temperatures than orthoclase. Partial substitution of potassium with sodium results in the formation of sanidine, which is stable at very high temperature or an-orthoclase (Table 3.6).

Table 3.5 Plagioclase series based on mixtures of albite and anorthite

Mineral phase	% Substitution of Ca^{2+} to Na^+	Example formula	Comments
Albite	0	$NaAlSi_3O_8$	Mineral name still albite even if Ca^{2+} substitutes up to 10% Na^+
Oligoclase	20	$(Na_{0.8}, Ca_{0.2}) AlSi_2O_8$	Mineral name still albite if Ca^{2+} substitutes between 10 and 30% of Na^+
Andesine	40	$(Na_{0.6}, Ca_{0.4}) AlSi_2O_8$	Mineral phase name still albite if Ca^{2+} substitutes between 30 and 50% of Na^+
Labradorite	60	$(Na_{0.4}, Ca_{0.6}) AlSi_2O_8$	Mineral phase name still albite if Ca^{2+} substitutes between 50 and 70% of Na^+
Bytownite	80	$(Na_{0.2}, Ca_{0.8}) AlSi_2O_8$	Mineral phase name still albite if Ca^{2+} substitutes between 70 and 90% of Na^+
Anorthite	100	$CaAl_2Si_2O_8$	Mineral name is anorthite even if Na^+ substitutes up to 10% Ca^{2+}

Table 3.6 K-feldspar series based on Na substitution

Mineral phase	Example formula	Crystal system
Orthoclase	$KAlSi_3O_8$	Monoclinic
Microcline	$KAlSi_3O_8$	Triclinic
Sanidine	$(K,Na)AlSi_3O_8$	Monoclinic
Anorthoclase	$(Na,K)AlSi_3O_8$	Triclinic

3.13 The Feldspar Solid Solution Using Rules of Ions Grouping

Given that the radius ratio of Al^{3+} is 0.38, this means that it can exist in a tetrahedral position in place of silicon and satisfies the basic rules of substitutions. The Al^{3+} cations substitute for the Si^{4+} cations in the tetrahedra resulting in positive charge deficiency. Only one or two out of every four tetrahedra may have a silicon cation replaced by aluminum cation, where more cations are very difficult to be accommodated by the crystal structure. The substituted tetrahedra require another cation or cations to balance the charge deficiency including K^{+}, Na^{+}, or Ca^{2+}. In the case of potassium feldspar, the charge deficiency due to the substitution of silicon cations by aluminum cations is balanced by insertion of a potassium (K^{+}) cation. In sodium feldspar, the charge deficiency is balanced by insertion of a sodium (Na^{+}) cation. The solid solution in feldspars is found to be a coupled ionic substitution, rather than the simple substitution.

At the same time, there are several other substitutions that can occur between potassium and sodium ions and sodium and calcium ions. The radius ratios for sodium and calcium cations are 0.73 and 0.71, respectively. So, the substitution occurring at higher temperature occurs between calcium and sodium and results in the plagioclase series. On the other hand, the substitution between potassium with radius ratio of 0.99 and sodium at higher temperature results in the formation of the alkali feldspar series.

3.14 The Basic Structural Units of Silicates

The chemical elements silicon and oxygen are found to make up 75% of the earth's crust, thus silicates are the most common minerals in the crust. When one atom of silicon and four atoms of oxygen combine in nature, the resulting configuration is a silica tetrahedron. Within the mineral structure, these silica tetrahedra link in various degrees to form a variety of silicate structure types.

Silicates are characterized by a complex chemical composition, although their basic structures have a simple ordered atomic arrangement. The radius ratio for a silicon cation to an oxygen anion is 0.29, which corresponds to a Si^{4+} cation coordinated by $4O^{2-}$ anions tetrahedral coordination. The four oxygen anions are found regularly arrayed around the central silicon cation. In order to neutralize the +4 charge of the Si^{4+} cation, one negative charge from each of the oxygen anions

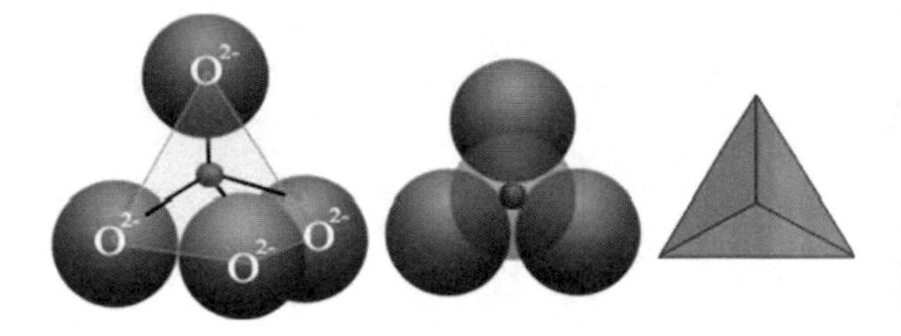

Fig. 3.8 A silica tetrahedron

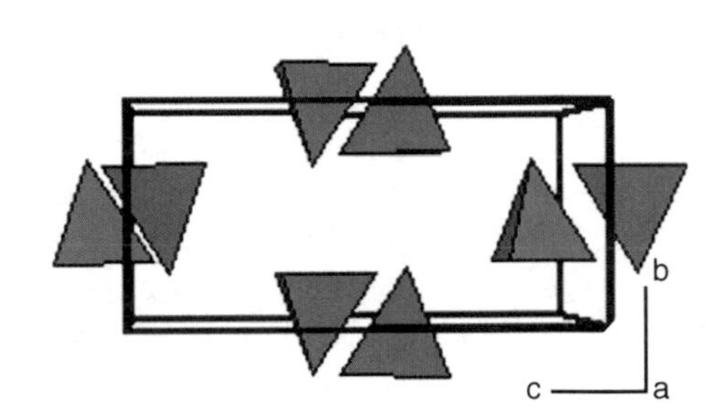

Fig. 3.9 Structure of forsterite

will bond with the Si^{4+} cation. This will leave each oxygen with a net charge of −1, resulting in a silica tetrahedron (SiO_4^{4-}). These silica tetrahedra form the basis of the silicate structure minerals (Fig. 3.8).

Since the Si^{4+} cation is a highly charged cation, it should be separated as far as possible from other Si^{4+} cations in the structure. So, when these silica tetrahedra are linked together, only corner oxygen anions of the tetrahedra will be shared between different tetrahedra. So, the silica tetrahedra can be linked in different compounds via several ways to provide the silicate with more open structures. This will yield several possibilities of different silicate groups. These groups include the so-called neosilicates, sorosilicate, inosilicates, cyclosilicates, phyllosilicates, and framework silicates.

3.14.1 Neosilicates (Single Tetrahedra)

The structure of this group is formed of isolated tetrahedra, and the corner oxygens in the silica tetrahedra are not shared. The basic structural unit is isolated silica tetrahedra with formula of SiO_4^{4-}. The oxygens are only able to share with other cations such as Mg^{2+}, Fe^{2+}, or Ca^{2+} in octahedral coordination. This assembly is being considered as an array of silicon tetrahedra with Mg^{2+} ions in the octahedral holes, where each oxygen is coordinated with one Si^{4+} and three Mg^{2+}. This group includes the olivine minerals (forsterite Mg_2SiO_4 and fayalite Fe_2SiO_4), the aluminosilicates such as kyanite, andalusite, sillimanite, and mullite. The structure of forsterite is shown in Fig. 3.9, where the oxygen anions are nearly arranged in a hexagonal close packed structure with Mg^{2+} in octahedral sites and Si^{4+} in tetrahedral sites.

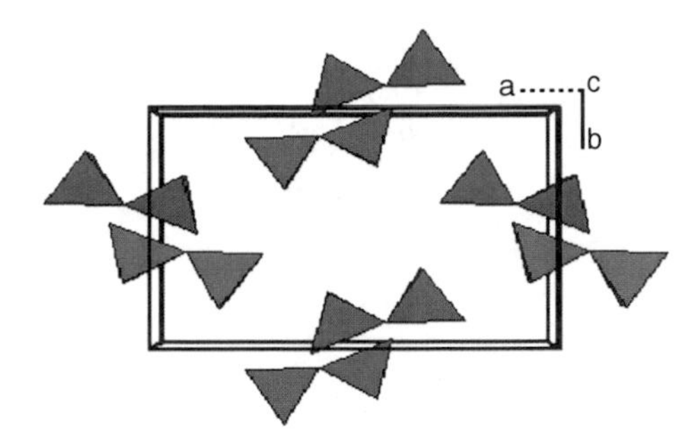

Fig. 3.10 Structure of sorosilicates

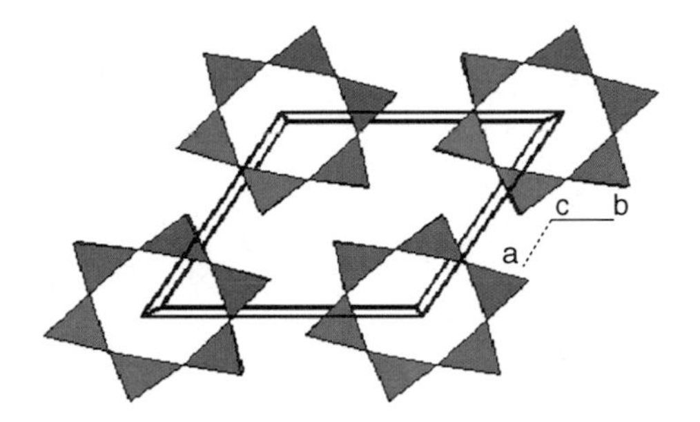

Fig. 3.11 Structure of cyclosilicates formed with six-membered rings (Beryl)

3.14.2 Sorosilicates (Double Tetrahedra)

When two tetrahedra are linked together by one of the corner oxygen these form the sorosilicate group as shown in Fig. 3.10. In this case, the basic structural unit is represented by the formula $Si_2O_7^{6-}$. A good example of a sorosilicate is the mineral hemimorphite – $Zn_4Si_2O_7(OH){\cdot}H_2O$.

3.14.3 Cyclosilicates (Ring Silicates)

Each tetrahedron is shared by two corner oxygen with two other tetrahedra and arranged in a ring to give the basic structural unit of the cyclosilicates. The ring structure can consist of six, five, four, or three-membered rings. The six-membered ring forms the structural unit $Si_6O_{18}^{12-}$, five-membered rings form the structural unit $Si_5O_{15}^{10-}$, the four-membered rings forms $Si_4O_{12}^{8-}$, and three-membered rings form the structural unit $Si_3O_9^{6-}$. A good example of the six-membered ring cyclosilicate is the mineral Beryl ($Be_3Al_2Si_6O_{18}$). The structure of Beryl is shown in Fig. 3.11. Each tetrahedron is linked at two corners to form isolated rings of silicate tetrahedra.

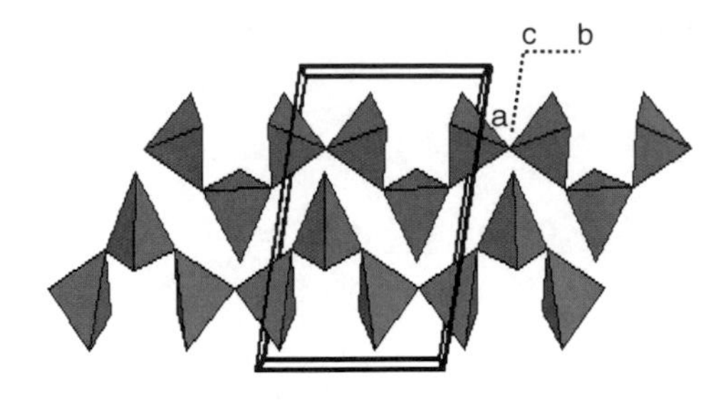

Fig. 3.12 Structure of a single chain silicate

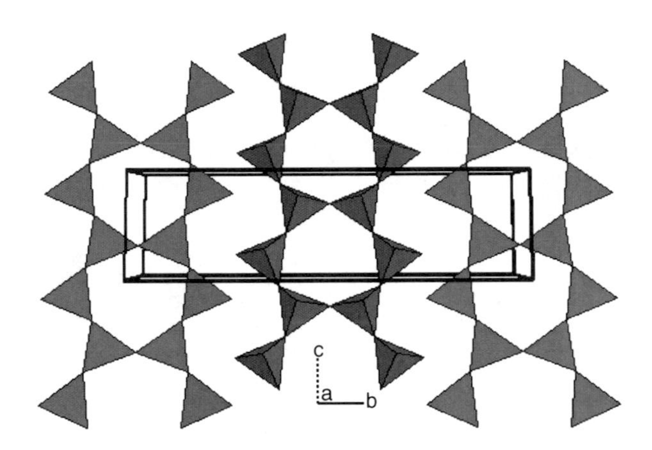

Fig. 3.13 Structure of the amphibole group of silicates

3.14.4 Inosilicates (Chain Structure Silicates)

In this group, each tetrahedron is shared by two corner oxygen with two other tetrahedra and arranged in either single chain or double chain silicates. Single chain silicates are represented by the basic structural unit $(SiO_3)_n^{n-}$. In this case, the basic structural unit may be $Si_2O_6^{4-}$ or SiO_3^{2-} as shown in Fig. 3.12 This group is the basis for the pyroxene group of minerals, like the orthopyroxenes (Mg,Fe)SiO_3 or the clinopyroxenes Ca(Mg,Fe)Si_2O_6 solid solutions. The pyroxene group includes enstatite, diopside, spodumene, and jadeite mineral phases.

Another arrangement involves two single chains linked together, so that each tetrahedron is shared by three of its oxygens to form a double chain silicate group with the basic structural unit being $Si_4O_{11}^{6-}$. The double chain group of minerals is represented by the amphibole group of minerals with the structure shown in Fig. 3.13.

3.14.5 Phyllosilicates (Sheet Structure Silicates)

In this group, each tetrahedron is linked at three corners to form the structure of parallel silicate sheets tetrahedral (Fig. 3.14). In this case, the basic structural

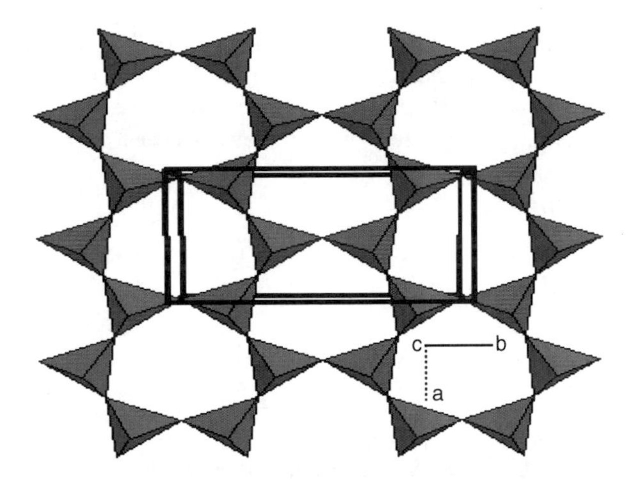

Fig. 3.14 Structure of phyllosilicates

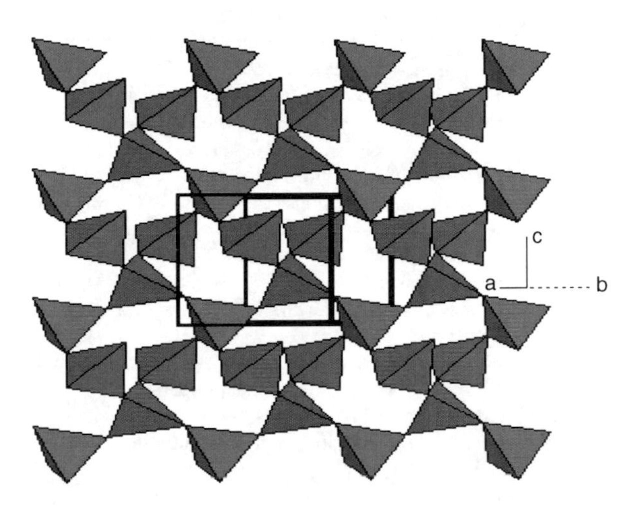

Fig. 3.15 Structure of a framework silicate

unit is $Si_2O_5^{2-}$. The micas, clay minerals, chlorite, talc, and serpentine minerals are all based on the phyllosilicate sheet structure. A good example is Biotite [$K(Mg,Fe)_3(AlSi_3)O_{10}(OH)_2$]. Note that in this structure Al is substituting for Si in one of the tetrahedral sites.

3.14.6 *Tectosilicates (Framework Silicates)*

When every tetrahedron is linked at each corner the structure forms a three-dimensional framework silicate (Fig. 3.15). All of the corner oxygens are shared with other SiO_4 tetrahedra and the basic structural unit will be SiO_2. The minerals quartz, cristobalite, and tridymite all are based on this structure. If some of the Si^{4+} ions are replaced by Al^{3+} then this produces a charge imbalance and allows for other ions to be found coordinated in different arrangements within the framework structure. The feldspars are formed of a framework formed of Al^{3+} replacing some of the Si^{4+} to make a framework with a net negative charge that is balanced by large ions in the interstitial positions, that is, albite, anorthite, orthoclase, and celsian in addition to the feldspathoid structures as nepheline mineral phase. The network structures are similar to cristobalite structure with the alkali or alkaline earth ions fitting into interstices.

Further Reading

Bioactive Glasses, Ceramics, Composites, Other Advanced Materials, Report Code: AVM054A, Analyst: Margareth Gagliardi. http://www.bccresearch.com/ (2006)

Boccaccini, A.R.: Assessment of brittleness in glass-ceramics and particulate glass matrix composites by indentation data. J. Mater. Sci. Lett. **15**, 1119–1121 (1996)

Burg, K.L., Porter, S., Kellam, J.F.: Biomaterial developments for bone tissue engineering. Biomaterials **21**(23), 2347–2359 (2000)

Cho, M., Lee, Y., Lim, B., Lim, Y.: Changes in optical properties of enamel porcelain after repeated external staining. J. Prosthet. Dent. **95**(6), 437–443 (2006)

Fishmann, G., Hench, L.: Bioceramics: Materials and Applications. American Ceramic Society, Westerville, OH, USA (1994)

Heffernan, M., Aquilino, S., Arnold, M., Haselton, D., Stanford, C., Vargas, M.: Relative translucency of six all-ceramic systems. Part I: Core materials. J. Prosthet. Dent. **88**(1), 4–9 (2002a)

Heffernan, M., Aquilino, S., Arnold, M., Haselton, D., Stanford, C., Vargas, M.: Relative translucency of six all-ceramic systems. Part II: Core and veneer materials. J. Prosthet. Dent. **88**(1), 10–15 (2002b)

Hench, L.L.: Bioceramics. J. Am. Ceram. Soc. **81**(7), 1705–1728 (1998)

Höland, W.: Biocompatible and bioactive glass-ceramics-state of the art and new directions. J. Non-Cryst. Solids **219**(1), 192–197 (1997)

Höland, W., Rheinberger, V., Apel, E., Van't Hoen, C., Höland, M., Dommann, A., Obrecht, M., Mauth, C., Graf-Hausner, U.: Clinical applications of glass-ceramics in dentistry. J. Mater. Sci. Mater. Med. **17**(11), 1037–1042 (2006)

Kelly, R.J.: Ceramics in restorative and prosthetic dentistry. Annu. Rev. Mater. Sci. **27**, 443–468 (1997)

Kima, D., Lee, J., Sung, R., Kim, S., Kim, H., Park, J.: Improvement of translucency in Al_2O_3 ceramics by two-step sintering technique. J. Eur. Ceram. Soc. **27**, 3629–3632 (2007)

Kingry, W.D., Bowen, H.K., Uhlmann, D.R.: Introduction to Ceramics, 2nd edn. Wiley, New York (1960)

Kingery W.D., Bown H. K., Uhlmann D. R.: Introduction to Ceramics, 2nd edition, J Wiley and Sons, Inc. (1976)

Lang, S., Starr, C.: Castable glass ceramics for veneer restorations. J. Prosthet. Dent. **67**(5), 590–594 (1992)

Lawn, B.R., Marshall, D.B.: Hardness, toughness, and brittleness by indentation analysis. J. Am. Ceram. Soc. **62**, 347–350 (1979)

Park, J., Lakes, R.: Biomaterials: An Introduction, 3rd edn. Springer, New York/London (2007)

Pauling, L.: Nature of the Chemical Bond, 3rd edn. Cornell University Press, Ithaca New York (1960)

Sehgal, J., Ito, S.: A new low-brittleness glass in the soda-lime-silica glass family. J Am. Ceram. Soc. **81**, 2485–2488 (1998)

Sehgal, J., Nakao, Y., Takahashi, H., Ito, S.: Brittleness of glasses by indentation. J Mater. Sci. Lett. **14**, 167–169 (1995)

Sukumaran, V., Bharadwaj, N.: Ceramics in dental applications. Trends Biomater. Artif. Organs **20**(1), 7–11 (2006)

Tsitrou, E.A., Northeast, S.E., van Noort, R.: Brittleness index of machinable dental materials and its relation to the marginal chipping factor. J. Dent. **35**(12), 897–902 (2007)

Part II
Manufacturing of Medical Glasses

Chapter 4
Formulation of Medical Glasses

The process of glass formulation is simply a process by which a glass is prepared with the desired properties for a certain application. The science of glass formulation is the knowledge that provides the possibility of producing distinctive formulations of value products for valued consumers. Creating the products starts by successfully combining the market view and the experience in formulation. The scientist skilled in glass formulation is able to design and develop glass formulations to meet specific needs for the application in mind.

In this chapter, we explain how to calculate glass chemical compositions in wt% or mol% and the proposed batch composition for glass precursors used to produce medical glass ceramics. Some of the developed glasses can be crystallized into medical glass ceramic products by heat treatment. Thus, we also consider on what basis we choose the chemical composition of the glass that will be utilized for producing glass ceramics for medical applications. We show a simple method of how to convert the molar glass chemical composition to the weight chemical composition and vice versa.

In addition, the chapter explains in brief the glass batch calculation for medical glasses and glass ceramics using different raw materials and the basis of the batch calculations. We also briefly throw light on how to reach the desired final glass composition, which will be melted to yield the final proposed glass composition.

4.1 Glass Chemical Compositions

The final chemical composition of the glass product needs to be controlled accurately. Therefore, the final composition of the glass should be as close as possible to the desired glass product chemical composition. This means that the glass chemical composition should give the final glass properties within criteria for the contents of the different glass components (usually oxides). These criteria should give the desired glass properties within a permitted variation in the content of each

E. El-Meliegy and R. van Noort, *Glasses and Glass Ceramics for Medical Applications*,
DOI 10.1007/978-1-4614-1228-1_4, © Springer Science+Business Media, LLC 2012

Table 4.1 Examples of the various starting glass chemical compositions in wt%

	wt%				
Oxides	Glass 1	Glass 2	Glass 3	Glass 4	Glass 5
SiO_2	35.16	40.42	34.16	27.86	25.03
BaO	48.51	50.70	52.36	48.05	49.32
MgO	4.72	8.88	9.18	4.68	8.64
B_2O_3	3.98	0.00	4.30	11.84	12.16
ZnO	7.63	0.00	0.00	7.56	4.85
Total	100.00	100.00	100.00	100.00	100.00

Table 4.2 Examples of the various starting glass chemical compositions in mol%

	Chemical composition of glasses (mol%)				
Oxides	Glass 1	Glass 2	Glass 3	Glass 4	Glass 5
SiO_2	50.0	55.0	45.00	40.00	35.00
BaO	27.0	27.0	27.00	27.00	27.00
MgO	10.0	18.0	18.00	10.00	18.00
B_2O_3	5.0	0.0	5.00	15.00	15.00
ZnO	8.0	0.0	0.00	8.00	0.00
Total	100.0	100.0	100.00	100.00	100.00

component. If the content or the type of one of the constituting oxides changes then the characteristics of the glass will definitely change.

The characteristics of glasses are known to be very sensitive to even very minor changes in the glass compositions. From the practical point of view, there are three main chemical compositions that can be calculated for glass preparation. The three compositions include the starting glass chemical composition in wt%, the glass chemical composition in mol%, and the batch composition.

In most practical work, we use only the chemical composition in wt% as a basis for the calculation of the batch composition that can be melted to provide the final desired glass chemical composition. In this chapter, we give a detailed description of how it is possible to calculate the different chemical compositions either in wt% or mol%.

All glasses are based on a starting chemical composition that is always listed in oxide wt% and sometimes in mol% for a certain purpose concerning the substitution of ions in the structure. Some glasses may contain metal fluorides such as MgF_2, AlF_3 or CaF_2, which are necessary as a precursor for the development of canasite glass ceramics or mica glass ceramics. Some examples of the various starting glass chemical compositions are shown in Tables 4.1 and 4.2.

4.2 The Glass Stoichiometry

Stoichiometry deals with the relative amount of reactants and products in a chemical reaction. In a balanced chemical reaction, the relations among quantities of reactants and products typically form a ratio of whole numbers. The term stoichiometry is also often used for the molar proportions of elements in stoichiometric compounds, which are then referred to as compositional stoichiometry.

We can consider the preparation of glass to be a chemical reaction. So we perform a melting of various reactants, usually reagent grade chemicals that yield the final desired glass chemical composition after complete melting and casting to produce glass. The glass stoichiometry is the quantitative relationship between the amounts of reactants and final reaction products. The stoichiometry is adjusted if the starting glass chemical composition (in the form of weight oxide chemical composition or its representative batch composition) yields the desired final glass chemical composition. But if the desired final chemical composition is not achieved, we have to study the factors leading to the compositional variation.

4.3 Factors Affecting the Glass Stoichiometry

In the laboratory, it is easy to control the chemical composition by using pure laboratory chemical reagents to satisfy the required glass chemical composition. Although it is easy for a skilled researcher to adjust the glass stoichiometry in the laboratory, there are still several factors that can affect the proposed matching of a chemical composition in laboratory work.

In the laboratory, some glass components, such as silica and zirconia, show no significant loss during glass melting. Any loss of such components is generally a result of errors in weighing the batch, loss during milling, or calculation of the wrong glass chemical formulation. This problem is readily avoided by ensuring the purity of the utilized raw materials and making sure that the balance is optimally calibrated. If the raw materials are not pure and contain impurities that are compatible with the proposed glass composition, it is still possible to satisfy the desired glass chemical composition. This can be dealt with by taking the presence of these impurities into consideration when calculating the glass batch composition.

Another factor that can affect the glass stoichiometry is the volatilization of low melting components during melting. The glass composition may lose some components during volatilization, a typical example being fluorine. The loss of fluorine may be as high as 25% during melting due to volatilization as in the case of melting a canasite glass at 1,400°C. To avoid excessive volatilization, melting can be carried out in covered crucibles. Alternatively, the projected loss is taken into account in the starting chemical composition. Careful control over fluorine content in canasite can reduce the loss to less than 10%.

A number of components that are subject to volatilization at glass melting temperatures that are significantly higher than the component melting temperature are listed in Table 4.3. Sodium oxide loss can be as high as 10% at higher melting temperatures. The loss can be quantified through the correlation of the starting chemical composition and the final glass chemical composition.

Other factor affecting the stoichiometry is the moisture content in the starting materials. Some raw materials such as carbonates are known to absorb moisture

Table 4.3 Melting temperatures of various low temperature glass components

Components	Na_2O	B_2O_3	P_2O_5	LiF	CaF_2	MgF_2
Melting, °C	1,132	480	569	848	1,402	1,263

from the atmosphere, a fact that can result in a significant change in the chemical composition. It should be appreciated that the moisture content entrapped in the structure of oxides is vital and needs to be taken into account in order to adjust the stoichiometry. For example, magnesium oxide, lanthanum oxide, and titanium oxide are known to absorb moisture from the atmosphere.

It is not possible to predict how much moisture might be entrapped in MgO, La_2O_3, or TiO_2. When these compounds are exposed to the atmosphere during weighing a glass batch, this makes it very difficult to adjust the final chemical composition. Also this moisture is not easily removed at low temperature drying. One way to avoid the problem is to heat the oxide at 600°C/2 h and then cool it, isolated from the atmosphere, in a desiccator just before using it. Then the samples need to be weighed out as fast as possible on removal from the desiccator to reach a chemical composition that is as close as possible to the stoichiometry.

4.4 Industrial Factors Affecting Glass Stoichiometry

In industry, the position is very complex compared to what happens in the research laboratories. The factors affecting the glass chemical composition are much more variable in industry. In addition to the previously discussed factors, other factors that need to be taken into account include the following:

1. Wrong starting chemical composition.
2. Wrong choice of the desired raw material, which may be incompletely dissociated during melting.
3. Noncalibrated weighing equipment.
4. Mechanical losses during batch processing.
5. Loss or gain due to reactions with the refractory walls.
6. Losses due to splitting the batch during melting.
7. Ejected particles during vigorous melting, which depends on the particle size.
8. Poor mixing of the batch.

To ensure that the final glass products have the required properties for the application in mind, the final chemical composition of the glass product therefore needs to be accurately controlled. This is done by ensuring that the glass final composition is as close as possible to the desired glass product chemical composition. This can be achieved through a careful choice and use of the proposed raw materials.

Another very important aspect is to understand the way in which the raw materials react or decompose during glass melting. It is always desirable to choose a base batch composition for a certain glass product appropriate to achieve the ultimate glass chemical composition. Having chosen the desired glass composition, the glass components can be selected from different raw material sources. It is very important to keep the glass constituents (usually oxides) close to the desired glass composition, irrespective of the different sources of the raw material one decides to use.

The compatibility of the raw materials chosen to satisfy the glass chemical composition should be checked. If one chooses a raw material that contains one or more of the constituents that is not represented in the glass, this will result in a different final product chemical composition. Thus, it would be undesirable to have, for example, Na_2O in the glass, when it is not represented in the glass chemical composition. Similarly, one should not use borax ($Na_2B_4O_7 \cdot 10H_2O$) as a source of boron oxide (B_2O_3) in a product that does not contain sodium oxide in its chemical composition, as each mole of borax contains 1 mole of Na_2O and 2 mole of B_2O_3. This means that the sodium oxide will enter in the structure of the glass product resulting in several changes in the glass characteristics, which will manifest as an increase in the thermal expansion, a decrease in the melting and softening temperature, and may also result in an increase in the chemical solubility of the glass.

4.5 Replacement of Oxygen by Fluorine in Glass Chemical Compositions

It is desirable in several medical glasses to replace oxygen by fluorine. The presence of the fluorine in glasses is generally reported as F_2 molecule in the analysis. Fluoride containing glasses already used in medical field include fluorophlogopite mica glasses. There are also fluoride containing glasses that are still at the stage of being investigated for future medical applications such as fluorcanasite glasses and fluorrichterite.

The fluorine is normally added to the batch in the form of fluorides such as NaF, CaF_2, or MgF_2. So, in case of medical glasses it will be more convenient if we calculate the excess oxygen as a result of fluorides dissociation. The dissociation of 1 mol of MgF_2 results in 1 mol of MgO and one molecule of F_2 is as follows:

$$MgF_2 \rightarrow MgO + F_2.$$

This example shows that the excess in oxygen content needed for every F_2 mole (38 g) will be half of oxygen mole (½O_2) which is equivalent to 16 g. This means that the excess in oxygen content for every 1 wt% of $F_2 = 1 \times 16/2 \times 19 = 0.42$ wt% of Oxygen. This can be calculated from the following formula.

$$\text{Excess in oxygen content} = \frac{F_2 \text{ content in wt\%} \times \text{atomic wt. of oxygen}}{\text{Molecular weight of } F_2}$$

So if for example a glass contains 4 wt% fluorine, this means that, it will gain excess oxygen $= 4 \times 0.42 = 1.64$ wt% from the atmosphere.

4.6 Information Needed for Glass Calculations

To start calculating the glass chemical composition first we need the starting glass chemical composition in wt% from which the raw materials batch will be calculated. Second, we need the chemical analysis of the chosen raw materials. Third, we need the molecular weights of the glass batch materials such as oxides, carbonates, hydroxides, phosphates or natural minerals. The molecular weights of ceramic oxides and fluorides that one is most likely to come across in the production of medical glasses are shown in Table 4.4.

Table 4.4 Molecular weights of medical ceramic oxides and fluorides

No.	Oxide or fluoride	Formula	Molecular weight
1	Aluminum fluoride	AlF_3	83.98
2	Aluminum oxide	Al_2O_3	101.96
3	Silver oxide	Ag_2O	231.74
4	Boron oxide	B_2O_3	68.02
5	Barium (II)oxide	BaO	153.33
6	Beryllium oxide	BeO	25.01
7	Calcium fluoride	CaF_2	78.08
8	Calcium oxide	CaO	56.08
9	Cerium oxide	CeO_2	172.11
10	Cobalt oxide	CoO	74.93
11	Chromium oxide	CrO_2	151.99
12	Cupper (II) oxide	CuO	79.55
13	Erbium oxide	Er_2O_3	382.52
14	Iron oxide	Fe_2O_3	159.69
15	Potassium oxide	K_2O	94
16	Lanthanum oxide	La_2O_3	325.81
17	Lithium fluoride	LiF_2	25.94
18	Lithium oxide	Li_2O	99.88
19	Magnesium fluoride	MgF_2	62.3
20	Magnesium oxide	MgO	40.3
21	Manganese oxide	MnO_2	86.94
22	Sodium oxide	Na_2O	62
23	Niobium pentoxide	Nb_2O_5	265.81
24	Nickel oxide	Ni_2O_3	74.69
25	Phosphorous pentoxide	P_2O_5	141.94
26	Antimony pentoxide	Sb_2O_5	291.52
27	Silicon dioxide	SiO_2	60.08
28	Tin oxide	SnO_2	150.71
29	Terbium oxide	Tb_2O_3	365.85
30	Titanium oxide	TiO_2	79.87
31	Vanadium pentoxide	V_2O_5	181.88
32	Yttrium oxide	Y_2O_3	225.81
33	Zinc oxide	ZnO	81.39
34	Zirconium oxide	ZrO_2	123.22

4.7 Calculation of Glass Chemical Compositions

Since this book deals with glasses that will be used as precursors for the preparation of medical glass ceramics, the focus here is on showing how to calculate medical glass chemical compositions and on what basis we can choose the proper chemical composition for glass ceramic products. The glass chemical compositions can be presented in two ways:

1. Chemical composition in wt%
2. Chemical composition in mol%

4.8 Glass Chemical Composition in wt% (Weight Composition)

The chemical composition of a glass listed in wt% is the one most commonly used. It lists the glass components in a way of expressing a number as a fraction of one hundred. It is used to express how large the weight of one component in the glass relative to another component. To show how to calculate the glass chemical composition in wt%, we present the following example of glass batch calculation.

Example 1

In this example, it is shown how to calculate a glass composition which, when it is heat treated at the respective temperature of nucleation and crystallization, yields a high strength, high toughness, and high hardness canasite glass ceramic.

4.8.1 Information Needed for Calculation

Canasite ($K_2Na_4Ca_5Si_{12}O_{30}F_4$) is a mineral phase, which can be synthesized from various oxides and calcium fluoride. The molecular weights of the canasite constituting oxides and calcium fluoride are shown in Table 4.5.

In case of medical glass ceramics, it is best to start with the stoichiometry of the mineral phase when determining the precursor glass compositions. Thus, we need to understand how to present each phase in terms of its constituting components, which are usually oxides. How to present the canasite chemical composition in the form of oxides together with CaF_2 is shown in Table 4.6.

Table 4.5 The molecular weights of the canasite constituting oxides

Component	CaO	Na_2O	K_2O	CaF_2	SiO_2
Mol. wt.	56	62	94	78	60

Table 4.6 The canasite chemical composition

Phase	Formula	Components				
Canasite stoichiometry	$K_2Na_4Ca_5Si_{12}O_{30}F_4$	3CaO	$2Na_2O$	K_2O	$2CaF_2$	$12SiO_2$
Number of moles	1	3	2	1	2	12
Relative weights	1,262	3×56	2×62	1×94	2×78	12×60

Table 4.7 The way of calculation of the chemical composition of a glass based on the stoichiometric canasite composition ($K_2Na_4Ca_5Si_{12}O_{30}F_4$)

	Molecular		Weight of components		Composition in wt%	
Components of canasite	weight in grams	Number of moles	Number of moles × mol. wt.	Component oxide weight	Calculations	wt%
CaO	56	3	3×56	168	(168×100)/1,262	13.3
Na_2O	62	2	2×62	124	(124×100)/1,262	9.8
K_2O	94	1	1×94	94	(94×100)/1,262	7.4
CaF_2	78	2	2×78	156	(156×100)/1262	12.4
SiO_2	60	12	12×60	720	(720×100)/1,262	57.1
Total				1,262		100.0

As the formula shows, the canasite is made up from various components that include 3 mol of CaO, 2 mol of Na_2O, 1 mol of K_2O, 2 mol of CaF_2, and finally 12 mol of SiO_2.

4.8.2 *Steps of Calculation of the Glass Chemical Composition in wt%*

To calculate the chemical composition in wt%, we have to undertake the following steps, which are also illustrated in Table 4.7.

1. List the components (oxides, fluorides, etc.) of canasite as in column 1.
2. List the molecular weight of each component in grams as in column 2.
3. From the canasite formula count the number of moles of each component as shown in column 3.
4. Calculate the total weight of the glass in grams as follows

$$\text{weight of oxide in the glass in grams} = \text{number of oxide in moles} \times \text{the oxide molecular weight}$$

5. Calculate the components in wt% as follows

$$\text{wt\% of the oxide in the glass} = (\text{number of component in moles} \times 100) / \text{total weight of all components in the glass in grams}$$

The total weight percentage of all the components should add up to 100% as shown in the final column.

4.9 Glass Chemical Composition in mol% (The Molar Composition)

4.9.1 Definition of a Mole

A mole of any substance consists of approximately 6.02214×10^{23} units of the substance. The value 6.02214×10^{23} is known as Avogadro's number. So we can define the mole as the amount of substance in grams, which contains approximately 6.02214×10^{23} of substance units.

The Avogadro's constant can be applied to any substance. It corresponds to the number of units (atoms or molecules or any other formula units) needed to make up a mass equal to the mass of the substance (atomic mass, molecular mass, mass in grams).

A mole of any substance can be measured by weighing out the mass of the substance, which contains 6.022×10^{23} substance units. For example, CaO has a molecular mass of 56 g/mol (40 g of calcium Ca + 16 g of oxygen), which contains 6.022×10^{23} formula units of CaO.

For every substance, there is a mass, which is equivalent to 1 mol of that substance. The mass needed for any substance can be calculated from the formula of the substance. Several examples of the mass equivalent to mole of materials used in medical glass ceramics are shown in Table 4.8.

So the number of substance units is always constant and equals 6.022×10^{23} units/mol. The weights vary because the different atoms making up different substances have different atomic masses as we have recorded previously.

In case of magnesium, the mole contains atoms as units, while in case of oxygen gas; the mole contains molecules as units. On the other hand, compounds such as

Table 4.8 The mass equivalent to 1 mol of oxides in medical glass

Substance	Formula	Molecular weight (g)	Units in 1 mole
Calcium	Ca	40	6.022×10^{23} atoms
Magnesium	Mg	24.3	6.022×10^{23} atoms
Aluminum	Al	26.98	6.022×10^{23} atoms
Oxygen	O_2	32	6.022×10^{23} molecules
Calcium oxide	CaO	56	6.022×10^{23} CaO formula units
Magnesium oxide	MgO	40.3	6.022×10^{23} MgO formula units
Phosphorous pentoxide	P_2O_5	141.9	6.022×10^{23} P_2O_5 formula units
Aluminum oxide	Al_2O_3	101.96	6.022×10^{23} Al_2O_3 formula units
Calcium carbonate	$CaCO_3$	100	6.022×10^{23} $CaCO_3$ formula units

Table 4.9 Method for calculating the stoichiometric canasite composition ($K_2Na_4Ca_5Si_{12}O_{30}F_4$) in mol%

Components of canasite	Component number of moles	Composition, mol% (Component number of moles × 100)/total	Composition, mol%
CaO	3.0	(3 × 100)/20	15.0
Na_2O	2.0	(2 × 100)/20	10.0
K_2O	1.0	(1 × 100)/20	5.0
CaF_2	2.0	(2 × 100)/20	10.0
SiO_2	12.0	(12 × 100)/20	60.0
Total	20.0		100.0

CaO are made up of two atoms including calcium and oxygen. These atoms are attached to each other such that the CaO behaves like a single unit rather than each of its individual calcium and oxygen atoms.

4.9.1.1 Why We Need to Calculate the Molar Chemical Composition (in mol%) of Glass

The chemical composition in mol% is important for many scientific reasons. It is used to prepare glasses which contain a given number of atoms. So it is possible to compare the behaviors and properties of these glasses keeping the same number of atoms. This means that there will be a greater chance to understand the different ionic substitutions in a glass such as replacing Na^+ by K^+ or Ca^{2+} by Mg^{2+} in the glass ceramic structure.

4.9.2 Steps of Calculation of the Chemical Composition in mol%

To calculate the chemical composition in mol% of canasite we have to take the following steps, also shown in Table 4.9:

1. List the components of canasite as in column 1.
2. From the canasite formula count the number of moles of each component that will provide the required weight in the canasite as shown in column 2.
3. Calculate the total number of moles of all the constituting components (oxides and fluorides, etc.) as shown in column 2.
4. Calculate the components in mol% as shown in column 3 using the following equation:

$$\text{mol\% of the component} = (\text{number of moles} \times 100)/\text{total number of moles of all the components}.$$

The total of the molar compositions of all of the glass components has to be 100% as shown in column 3.

Table 4.10 Conversion of the chemical composition of canasite glass in mol% to the chemical composition in wt%

Components of canasite	Composition, mol%	Molecular weight, g	Weight of each component, g	Composition, wt%	
				Calculations	wt%
CaO	15.0	56.0	15×56=840	(840×100)/6,310	13.3
Na_2O	10.0	62.0	10×62=620	(620×100)/6,310	9.8
K_2O	5.0	94.0	5×84=420	(470×100)/6,310	7.4
CaF_2	10.0	78.0	10×78=780	(780×100)/6,310	12.4
SiO_2	60.0	60.0	60×60=3,600	(3,600×100)/6,310	57.1
Total	100.0		6310		100

4.10 Conversion of Molar Composition (mol%) to Weight Composition (wt%)

Let us convert the canasite glass chemical composition in mol% to the chemical composition in wt%, so we can calculate the batch composition and then the weight of the batch to be used for melting.

The molar composition in mol% is converted to the chemical composition in wt% following the sequence of steps as shown in Table 4.10:

1. List the components of canasite as in column 1.
2. List the chemical composition in mol% as shown in column 2.
3. List the molecular weight of each component as shown in column 3.
4. Calculate the weight in grams of each component as shown in column 4 from the following:

$$\text{The weight of each component in grams} = \text{Component in mol\%} \times \text{the molecular weight of that component.}$$

5. Calculate the components in wt% as shown in column 5 using the following equation:

$$\text{wt\% of the component} = (\text{weight of the component in grams} \times 100) / \text{total weight of all the components}$$

The total wt% of all the components has to be 100% as shown in column 5.

4.11 Conversion of Weight Composition (wt%) to Molar Composition (mol%)

The composition in wt% is converted to the chemical composition in mol% following the sequence of steps as shown in Table 4.11:

1. List the components of canasite in wt% as in column 2.
2. List the molecular weights of components as in column 3.

Table 4.11 Conversion of the chemical composition of canasite glass in wt% to the chemical composition in mol%

Components of canasite	Composition in wt%	Mol. wt.	wt%/mol. wt.	Number of moles	Composition, mol% Calculations	mol%
CaO	13.3	56	13.3/56	0.238	(0.238/1.585)×100	15.0
Na_2O	9.8	62	9.8/62	0.158	(0.158/1.585)×100	10.0
K_2O	7.4	94	7.4/94	0.079	(0.079/1.585)×100	5.0
CaF_2	12.4	78	12.4/78	0.159	(0.159/1.585)×100	10.0
SiO_2	57.1	60	57.1/60	0.952	(0.952/1.585)×100	60.0
Total	100.0			1.585		100.0

3. Calculate the number of moles as in columns 4 and 5 from

   number of moles = component wt%/molecular weight

4. Calculate the mol% as shown in columns 6 and 7 as follows

   The component in Mol% = (number of component's moles/the total number of moles of all components) × 100

   The total mol% of all the components has to be 100% as shown in column 7.

4.12 Calculation of the Glass Chemical Composition of K-Fluorrichterite

Now let us try to prepare a glass that crystallizes to form fluorrichterite glass ceramics for medical applications. The fluorrichterite ($KNaCaMg_5Si_8O_{22}F_2$) is a mineral phase that can be synthesized from various oxides and magnesium fluoride. The calculation of the chemical composition in wt% of fluorrichterite is a little bit more complex than that of the canasite. In these calculations, we use different sources of fluorine including MgF_2 and CaF_2.

4.12.1 Information Needed for Calculation

As the fluorrichterite ($KNaCaMg_5Si_8O_{22}F_2$) is a mineral phase that can be synthesized from various oxides and calcium fluoride or MgF_2, it is essential understand how to present any phase in its constituting components, which is usually oxides. Identifying the fluorrichterite chemical components first requires a balanced equation. The equation can be made using various materials. For example, we can use either MgF_2 as in (4.1) or CaF_2 as shown in (4.2).

$$2KNaCaMg_5Si_8O_{22}F_2 = K_2O \cdot Na_2O \cdot 8MgO \cdot 2MgF_2 \cdot 2CaO \cdot 16SiO_2 \quad (4.1)$$

$$2KNaCaMg_5Si_8O_{22}F_2 = K_2O \cdot Na_2O \cdot 10MgO \cdot 2CaF_2 \cdot 16SiO_2 \quad (4.2)$$

Table 4.12 The fluorrichterite mineral components using MgF_2 as a source of fluorine

Phase	Formula	Components					
Fluorrichterite	$2KNaCaMg_5Si_8O_{22}F_2$	K_2O	Na_2O	$8MgO$	$2MgF_2$	$2CaO$	$16SiO_2$
Number of moles	2	1	1	8	2	2	16
Relative weights	1,675	94	62	322.4	124.6	112	960

Table 4.13 Calculation of the chemical composition of a fluorrichterite glass in wt% using MgF_2 as a source of fluorine

			Weight of components		Composition, wt%	
Components of fluorrichterite	Component number of moles	Mol. wt.	Number of moles × mol. wt.	Component weight	Calculations	wt%
CaO	2	56.0	2 × 56	112.0	(112 × 100)/1,675	6.7
Na_2O	1	62.0	1 × 62	62.0	(62 × 100)/1,675	3.7
MgO	8	40.3	8 × 40.3	322.4	(322.4 × 100)/1,675	19.2
K_2O	1	94.0	1 × 94	94.0	(94 × 100)/1,675	5.6
MgF_2	2	62.3	2 × 62.3	124.6	(124.6 × 100)/1,675	7.4
SiO_2	16	60.0	16 × 60	960.0	(960 × 100)/1,675	57.3
Total				1,675		100.0

Thus, two moles of fluorrichterite mineral can be presented in the form of oxides together with MgF_2 following (4.1) as shown in Table 4.12:

4.12.1.1 Steps of Calculation of the Glass Chemical Composition in wt% Using MgF_2 as a Source of Fluorine

To calculate the chemical composition in wt% the sequence of steps required are shown in Table 4.13.

1. List the components of fluorrichterite as in column 1.
2. From the fluorrichterite formula count the number of moles of each component that will provide its needed weight in the fluorrichterite glass as shown in column 2.
3. Calculate the weight of each component as in column 3 from the following equation:

   weight of component = number of moles × molecular weight of the component

4. Calculate the components in wt% using the following equation:

   wt% of the component = (the weight of component × 100)/total weight of all the components

Table 4.14 Calculation of the chemical composition of a fluorrichterite glass in wt% using CaF_2 as a source of fluorine

Components of canasite	Component number of moles	Mol. wt.	Weight of components: Number of moles × mol. wt.	Component weight	Composition, wt%: Calculations	wt%
Na_2O	1	62.0	1×62	62.0	(62×100)/1,675.16	3.7
MgO	10	40.3	10×40.3	403	(403.4×100)/1,675.16	24.1
K_2O	1	94.0	1×94	94.0	(94×100)/1,675.16	5.6
CaF_2	2	78.08	2×78.08	156.16	(156.16×100)/1,675.16	9.3
SiO_2	16	60.0	16×60	960.0	(960×100)/1,675.16	57.3
Total				1,675.16		100.0

Table 4.15 Method of calculating the chemical composition of fluorrichterite in mol% *using MgF2 as a source for fluorine*

Components of canasite	Component number of moles	Composition, mol%: (Component number of moles × 100)/total	Composition, mol%
CaO	2	2×100/30	6.7
Na_2O	1	1×100/30	3.3
MgO	8	8×100/30	26.7
K_2O	1	1×100/30	3.3
MgF_2	2	2×100/30	6.7
SiO_2	16	16×100/30	53.3
Total	30		100

The total weight percentage of all the components has to be 100% as shown in column 7.

Fluorrichterite can be calculated in wt% using CaF2 as a fluorine source as shown in Table 4.14.

4.12.2 Steps of Calculation of the Glass Chemical Composition in mol%

Let us calculate the molar composition of fluorrichterite in mol% *using MgF_2 as a source for fluorine*. To calculate the chemical composition in mol%, we have to do the following as shown in Table 4.15.

1. List the components of fluorrichterite as in column 1.
2. From the fluorrichterite formula count the number of moles of each component that will provide its needed weight in the fluorrichterite as shown in column 2.

Table 4.16 Method of calculating the chemical composition of fluorrichterite in mol% using CaF_2 as a source for fluorine

Components of canasite	Component number of moles	Composition, mol% (Component number of moles × 100)/total	Composition, mol%
Na_2O	1	$1 \times 100/30$	3.3
MgO	10	$10 \times 100/30$	33.3
K_2O	1	$1 \times 100/30$	3.3
CaF_2	2	$2 \times 100/30$	6.7
SiO_2	16	$16 \times 100/30$	53.3
Total	30		100

3. Calculate the total number of moles of all the constituting oxides and fluorides components as shown in column 2.
4. Calculate the components in mol% as shown in column 3 using the equation:

$$\text{mol\% of the component} = (\text{number of moles} \times 100) / \text{total number of moles of all the components}$$

The total of the molar compositions of all of the glass components has to be 100% as shown in column 4.

The molar composition of k-fluorrichterite in mol% can also be calculated *using CaF2 as a source for fluorine* as shown in Table 4.16.

4.13 Conversion of Molar Composition to Weight Composition

The composition of fluorrichterite in mol% is converted into the chemical composition in wt% following the sequence of steps as shown in Table 4.17.

1. List the components of fluorrichterite as in column 1.
2. List the chemical composition in mol% as shown in column 2.
3. List the molecular weight of each component as shown in column 3.

Calculate the weight in grams of each component as shown in column 4 from the following:

The weight of each component in grams = component in mol% × the molecular weight of that component.

Table 4.17 Conversion of the chemical composition of fluorrichterite glass in mol% to its chemical composition in wt%

Components of canasite	Composition in mol%	Mol. wt., g	Weight of each component, g	Composition, mol% Calculations	wt%
CaO	6.7	56.00	373.3	(373.3 × 100)/5,583.3	6.7
Na_2O	3.3	62.00	206.7	(206.7 × 100)/5,583.3	3.7
MgO	26.7	40.30	1,074.7	(1,074.7 × 100)/5,583.3	19.2
K_2O	3.3	94.00	313.3	(313.3 × 100)/5,583.3	5.6
MgF_2	6.7	62.30	415.3	(415.3 × 100)/5,583.3	7.4
SiO_2	53.3	60.00	3,200.0	(3,200 × 100)/5,583.3	57.3
Total	100.0		5,583.3		100.0

4. Calculate the components in wt% as shown in column 5 using the following equation:

$$\text{wt\% of the component} = (\text{weight of the component in grams} \times 100) / \text{total weight of all the components}$$

The total mol% of all the components has to add up to 100% as shown in column 6.

4.14 Templates for Conversion of Glass of Molar Composition to Weight Composition

Table 4.18 provides a concise account of the various steps involved in the calculation of the different types of chemical analysis of a glass and the conversion between different presentations of the canasite. Table 4.19 shows the conversion steps for the chemical analysis of fluorrichterite.

4.15 Glass Batch Calculations

The glass batch is the mixture of raw materials that, when it is melted, produces a glass product with the desired chemical composition. The calculation of a batch formula is a significant target of glass laboratories and other industries that have to melt the batch to produce finished products, especially where tolerances are small as in the field medical glass ceramics. Errors in formula calculation are hard to identify and yet they can have a noticeable effect on the finished product quality. Mass-scale defects of finished products, if they are related to calculation errors, are usually attributed to the quality of batch proportioning and mixing.

Table 4.18 Brief description of how to calculate and switch between different types of chemical compositions of canasite

Oxide	Number of moles	mol%	Mol. wt.	Weight	wt%	Number of moles	mol%
CaO	3	15.0	56.0	840.0	13.7	0.2438	15.0
Na_2O	2	10.0	62.0	620.0	10.1	0.1625	10.0
K_2O	1	5.0	94.0	470.0	7.6	0.0813	5.0
CaF_2	2	10.0	62.3	623.0	10.1	0.1625	10.0
SiO_2	12	60.0	60.0	3,600.0	58.5	0.9751	60.0
Total	20	100.0		6,153.0	100.0	1.6252	100.0

Table 4.19 Template showing how to calculate and switch between different types of chemical compositions of a fluorrichterite glass

Oxide	Number of moles	mol%	Mol. wt.	Weight	wt%	Number of moles calc. from wt%	mol%
CaO	2	6.7	56.0	373.3	6.7	0.1194	6.7
Na_2O	1	3.3	62.0	206.7	3.7	0.1194	3.3
MgO	8	26.7	40.3	1,074.7	19.2	0.4776	26.7
K_2O	1	3.3	94.0	313.3	5.6	0.0597	3.3
MgF_2	2	6.7	62.3	415.3	7.4	0.1194	6.7
SiO_2	16	53.3	60.0	3,200.0	57.3	0.9552	53.3
Total	30	100.0		5,583.3	100.0	1.791	100.0

The current examples are chosen to illustrate the rules of glass batch calculations. In these examples, several assumptions must be taken into consideration during the batch calculations. The model of calculation assumes that

1. The raw materials are pure, dry, and their chemical analyses are consistent.
2. The raw materials react and decompose to completion during melting steps.
3. There are no oxide losses from the melt by volatilization.
4. There is no gain or loss due to reaction with refractory materials.
5. There is no gain due to the reaction with the furnace atmosphere.

4.16 Reaction and Decomposition of Raw Materials

4.16.1 Metal Carbonates

To understand the glass batch calculation and the displayed examples, we need to first understand the ways in which the raw materials can react and decompose during glass melting. As we are dealing with medical glasses and glass ceramics, so all the raw materials used in this example will be chosen from those used in manufacturing medical ceramics.

In the case of medical glasses and in turn glass ceramics, pure chemical reagent grade precursors are used. Carbonates are a main raw materials used to supply several oxide constituents in the glass composition such as Na_2O, K_2O, BaO, MgO, Li_2O, CaO.

Carbonates are assumed to decompose during the stages of melting reactions to yield oxide and carbon dioxide. All of the oxide content is assumed to enter in the structure of glass while the carbon dioxide is lost to the atmosphere. Some raw materials yield only one oxide to enter in the glass structure such as the following types:

Sodium carbonate (soda ash)	$Na_2CO_3 = Na_2O + CO_2$
Calcium carbonate (limestone)	$CaCO_3 = CaO + CO_2$
Magnesium carbonate (magnesite)	$MgCO_3 = MgO + CO_2$
Lithium carbonate	$Li_2CO_3 = Li_2O + CO_2$

Other carbonates can act as a source of two oxides to enter in the glass structure, e.g.,

Calcium magnesium carbonate (Dolomite) $CaMg(CO_3)_2 = CaO + MgO + 2CO_2$

4.16.2 *Metal Hydroxides*

Hydroxides can also be used as raw materials to satisfy several glass component oxides. Hydroxides are assumed to decompose during the stages of melting reactions to yield the oxide and water. All of the oxide content is assumed to enter in the structure of glass while the water is lost to the atmosphere.

Sodium hydroxide	$2NaOH = Na_2O + H_2O$
Aluminum hydroxide	$Al(OH)_3 = Al_2O_3 + H_2O$
Magnesium hydroxide	$Mg(OH)_2 = MgO + H_2O$
Potassium hydroxide	$2KOH = K_2O + H_2O$

4.16.3 *Borax and Boric Acid*

Borax is assumed to decompose completely to yield two oxides including sodium oxide and boric oxide that will enter in the structure of glass. On the other hand, boric acid is assumed to decompose to provide boric oxide that will enter in the composition of the glass.

Borax	$Na_2B_4O_7{\cdot}10H_2O = Na_2O + 2B_2O_3 + 10H_2O$
Boric acid	$2H_3BO_3 = B_2O_3 + 3H_2O$

4.17 Method of Calculation Glass Batch Compositions

Step 1

Let us calculate the batch composition for a glass with a chemical composition as listed in wt% in Table 4.20.

This is a simple example of the glass chemical composition. If you look in the constituting oxides, you will see these oxides including SiO_2, BaO, and MgO. In fact one can use any material that will be assumed to decompose and give the respective oxide percentage in the glass.

Step 2

Let us suggest three raw materials that can provide the above oxides, their molecular weights, and their formulas. We will source SiO_2 as silica. BaO will be sourced using barium carbonate, which decomposes during glass melting to give BaO and CO_2. Also, magnesium carbonate will be used to source MgO, while CO_2 evolves. The oxides, sourcing raw materials and their molecular weights are listed in Table 4.21.

Step 3

The next step is to determine how much barium carbonate will be required to provide the necessary amount of barium oxide in the glass and how much magnesium carbonate will be required to produce the necessary quantity of MgO in the glass. The answer will be easy if we determine the relative molecular weights of each of the required oxide and the raw material used.

It is essential to remember to balance the equations to calculate the right relative molecular weights. Now the relative molecular weights of BaO and $BaCO_3$ are shown through the decomposition reaction of barium carbonate from the following equation:

Reaction	$BaCO_3$	$\longrightarrow$	BaO	CO_2
Relative molecular weights	197.3		153.3	44.0

Table 4.20 Chemical composition of a glass listed in wt%

	Chemical composition in wt%			
Oxides	SiO_2	BaO	MgO	Total
wt%	40.4	50.7	8.9	100

Table 4.21 The oxides, sourcing raw materials and the molecular weights

Oxide	Oxide wt%	Molecular wt. of oxide	Raw material	Molecular wt. of raw material
SiO_2	40.4	60.0	Silica (SiO_2)	60.0
BaO	50.7	153.3	Ba-carbonate ($BaCO_3$)	197.3
MgO	8.9	40.3	Mg-carbonate ($MgCO_3$)	84.3

Table 4.22 The batch calculation template for a simple glass

Oxide	Oxide wt%	Raw material	Factor	Calculation	Batch, g
SiO_2	40.4	Silica	1.00	40.4×1	40.4
BaO	50.7	Barium carbonate	1.287	1.287×50.7	65.25
MgO	8.9	Magnesium carbonate	2.09	8.90×2.09	18.60

The relative molecular weights of MgO and $MgCO_3$ is shown through the decomposition reaction of magnesium carbonate as shown in the following equation

Reaction	$MgCO_3$	$\longrightarrow$	MgO	CO_2
Relative molecular weights	84.3		40.3	44.0

Step 4

Calculation of the amount of each raw material needed to provide a given weight of oxide. A factor can be calculated for each oxide from its raw material by dividing the molecular weight of the raw material by the molecular weight of the oxide. From the following equation, it is clear that the weight of barium carbonate needed to source 1 g of BaO equals:

$$\text{Factor of BaO} = \frac{\text{Molecular weight of BaCO}_3}{\text{Molecular weight of BaO}} = \frac{197.3}{153.3} = 1.287$$

$$\text{Factor of MgO} = \frac{\text{Molecular weight of MgCO}_3}{\text{Molecular weight of MgO}} = \frac{84.3}{40.3} = 2.092$$

On the other hand, SiO_2 will be added as silica so we do not need to calculate a factor. The amount of SiO_2 will be provided by simply adding 40.4 g of silica. So the factor of SiO_2 to be sourced from quartz silica is 1 assuming that the quartz silica is pure.

The batch composition can be calculated by multiplying the content of each oxide in the glass by the corresponding factor. The batch calculation of the previous chemical composition of the glass is shown in Table 4.22.

4.18 Batch Calculation for Stoichiometric Canasite Glass

The canasite ($K_2Na_4Ca_5Si_{12}O_{30}F_4$) is a mineral phase, which can be synthesized from various oxides and calcium fluoride. The batch calculation will pass through several steps as shown in Table 4.23.

1. List the chemical composition of the glass as shown in column 2.
2. List the molecular weights of glass components as shown in column 3.

Table 4.23 The batch calculation steps for a canasite glass

Oxide	wt%	Mol. wt. of oxide	Raw material	Mol. wt. of RM	Factor calculation	Factor	Batch calculation	Batch
CaO	13.30	56	$CaCO_3$	84	84/56	1.50	13.3 × 1.5	19.95
Na_2O	9.80	62	Na_2CO_3	106	106/62	1.71	9.8 × 1.71	16.75
K_2O	7.40	94	K_2CO_3	138	138/94	1.47	7.4 × 1.47	10.86
CaF_2	12.40	78	CaF_2	78	78/78	1.00	12.4 × 1	12.40
SiO_2	57.10	60	SiO_2	60	60/60	1.00	57.1 × 2	57.10
Total	100							

3. List the different utilized raw materials as shown in column 4.
4. List the molecular weights of the raw materials as shown in column 5.
5. Calculate the factor as in column 7.

   Factor = molecular weight of raw material/molecular weight of the component.

6. Calculate how much of the raw materials is needed to provide the oxide content in the glass.

   Materials needed to satisfy the oxide content in the glass = Oxide content in wt% × factor.

7. List the batch composition as in column 7.

4.19 Batch Calculation for Stoichiometric Fluorrichterite Glass

The fluorrichterite ($KNaCaMg_5Si_8O_{22}F_2$) is a mineral phase, which can be synthesized from various oxides and magnesium fluoride. The batch calculation will pass through several steps as shown in Table 4.24.

1. List the chemical composition of the glass as shown in column 2.
2. List the molecular weights of glass components as shown in column 3.
3. List the different utilized raw materials as shown in column 4.
4. List the molecular weights of the raw materials as shown in column 5.
5. Calculate the factor as in column 6.

   Factor = molecular weight of raw material/molecular weight of the component.

6. Calculate how much of the raw materials is needed to provide the oxide content in the glass.

   Materials needed to satisfy the oxide content in the glass = oxide content in wt% × factor.

7. List the batch composition as in column 7.

Table 4.24 The batch calculation steps for a fluorrichterite glass

Oxide	wt%	Oxide mol. wt.	Raw material	Mol. wt. of raw material	Factor	Batch composition
CaO	6.7	56	$CaCO_3$	84	1.50	10.05
Na_2O	3.7	62	Na_2CO_3	106	1.71	6.33
MgO	5.6	40.3	$MgCO_3$	84.3	2.09	11.71
K_2O	19.2	94	K_2CO_3	138	1.47	28.19
CaF_2	7.4	78	CaF_2	78	1.00	7.40
SiO_2	57.3	60	SiO_2	60	1.00	57.30
Total	100					

Further Reading

Andrews, A.I.: Enamels: The Preparation, Application, and Properties of Vitreous Enamels. The Twin City Printing, Champaign, IL (1935)

Hearh, A.: A Handbook of Ceramic Calculations. Stoke On Trent, Webberley Limited (1937)

Omar, A.A., Hamzawy, E.M.A., Farag, M.M.: Crystallization of fluorcanasite–fluorrichterite glasses. Ceram. Int. **35**, 301–307 (2009)

Parmelee, C.: In: Svec, J. (ed) Ceramic glazes, 3rd edition, completely revised by Cameron G. Harman. CBI Publishing, Boston, USA (1993)

Samsonov, G.V.: The Oxide Handbook, 2nd edn, p. 463. IFI/Plenum, New York (1982)

Chapter 5
Theoretical Estimation of Glass Properties

5.1 Importance of Estimation

The theoretical estimation of glass properties is very important for glass scientists, students, and workers to predict glass properties of interest and glass behavior under certain conditions before carrying out the experimental investigation. The estimation of properties saves time, material, financial, and environmental resources during the experimental work. It enables us to more or less predict or estimate the desired value of the property.

An example of this sort of approach to the design of a glass is the development of a ceramic coating for a base metal alloy. On coating a chromium nickel molybdenum alloy with a low fusion veneering glass ceramic, the thermal expansion coefficient must be adjusted to be between 13 and 15×10^{-6}/°C. A veneering glass ceramic with a 10% content of leucite crystals dispersed in glass matrix is predicted to produce a glass with a thermal expansion coefficient that is compatible with the alloy. With this estimation it is possible to match the thermal expansion coefficient of the veneering layer with that of the alloy by making few experiments with minimum experimental work, whereas the alternative would have been to produce hundreds of compositions and of course experiments to prepare a suitable veneering ceramic.

5.2 Additives Law

The properties of glassy materials can undergo significant changes with a change in composition and this includes changes in specific gravity, density, hardness, elasticity, refractive index, thermal expansion coefficient, and mechanical properties. Additivity is used to express the relationships between properties and chemical composition of glasses. The additive method was originally developed to give an approximate prediction of some glass properties based on their chemical compositions. The primary criterion for the success of the additive method is the close fit with the calculated values of the properties and the experimental values of

E. El-Meliegy and R. van Noort, *Glasses and Glass Ceramics for Medical Applications*, DOI 10.1007/978-1-4614-1228-1_5, © Springer Science+Business Media, LLC 2012

properties and it is now possible, with a considerable degree of confidence, to calculate many of the properties of a glass from its chemical composition.

The theoretical calculations are based on the additive factors that express the effect of a unit amount of oxide on the property of the glass. The factors are generally related to a unit amount of oxide in the glass expressed in wt% or mol%. The calculations consider that the glass is a mixture of oxides and sometimes contains fluorides such as MgF_2 or CaF_2. The glass must not be seen simply as a mixture of oxides, to the contrary, the glass is a solid solution of different oxides exhibiting various cation and anion substitutions. Thus, the properties of oxides in the glass are different from those of the isolated oxides. Therefore, we need to know the additive factors to imagine how much each oxide will affect a certain property of the glass. Where there are differences between the predicted and experimental values of any property, the reason would be attributed to the grouping of ions in the glass structure, the possibility of an ion existing in different coordination numbers and the way it was substituted and linked in the glass.

The grouping of ions in the glass structure is very important as it affects the accuracy of the additive factors. Cations may have different coordination numbers in the glass depending on their relationships with the rest of the glass composition. For example, aluminum and boron cations can coordinate in the structure with different coordination numbers ($^{[4]}$Al and $^{[6]}$Al or $^{[3]}$B and $^{[4]}$B) to create various ionic groupings, which also exhibit different glass physical properties. Also, it is not necessarily for the same cation connected with various numbers of the surrounding oxygen anions to always have different additive factors.

Normally, we calculate any property of an oxide in a trial to reach the value as close as to the desired value of that property from the starting chemical composition. The calculation of a certain property of a glass is based on the following additive relationship:

$$K = P_A X_A + P_B X_B + P_C X_C$$

where K is the specific property value to be calculated; X_A, X_B, and X_C are the weight percentage of each glass oxide components A, B, C; and P_A, P_B, and P_C are their specific property.

In some glasses, the accuracy of such calculations is very good, but in other glasses only an approximation is achievable. The degree of accuracy is normally related to the relationship between different oxides and their coordination in glasses, where some elements may influence the consequences of other elements. This fact may explain why the calculations of different authors may be relatively different. Thus, it is important to be aware of the major limitations using these formulae that need to be taken into consideration when performing the calculations. These limitations include the following points:

1. The calculations of properties from the glass chemical composition are approximate at best, even if the accuracy of the calculation for some glasses is very good.
2. The specific outcome for any oxide may be affected by the rest of the composition or the presence of other oxides or the constituting components in the composition.

The annealing conditions of the glass also affect the calculations.

The accuracy of the additive method of calculating glass properties thus depends on many factors, which may confound the prediction. Since this book is concerned with the calculations of medical glasses, we will not go further in exploring complex mathematical calculations. We will simply refer to how each glass property can be calculated, what factors affect the calculation of each property and give clear examples of how to calculate that property for medical glasses.

5.3 Calculation of Glass Density

The density of a glass responds sensitively to variations in glass compositions than any other physical property of a glass. Thus, density measurements are used routinely as a sensitive check to identify any chemical composition changes during the glass manufacturing.

The density of a homogeneous substance, for which the glass is a good example, is defined as the weight related to unit volume and expressed by the ratio of mass m and the volume V. The density of glass is generally measured at room temperature of 20°C from the following equation:

$$\text{Density}, \rho = m/V = \text{g/cm}^3.$$

The density will decrease gradually with increasing temperature above 20°C in the range below T_g as the density is dependent on the coefficient of thermal expansion. The density of a glass at a certain temperature can be derived from the following equation:

$$\rho_t = \rho_{20}/[1 - 3\alpha(t - 20)],$$

where ρ_t is the density at temperature t, α is the linear coefficient thermal expansion, ρ_{20} is the density at 20°C

Most glasses have higher densities than their corresponding crystalline oxides, such that the calculated densities will be higher than the measured ones. The volume of the melted glass is smaller than that of the sum of the volumes of the individual oxides. Melting brings about the densification of the components resulting in the formation of silicates with a new special arrangement of ions. The ions in the glass will be arranged more densely in the glass structure than in the molecules of the constituting oxides. Thus, the densities of almost all of the oxides incorporated in a glass are usually higher than the densities of their free oxides.

The additive method is used to calculate the density of a glass from its starting chemical composition. The method uses the density of oxides bound in the glass related to the fraction of the corresponding oxide.

Table 5.1 The densities of a range of free oxides and their corresponding densities when bound in the glass as adapted from Volf (1988)

No.	Oxides	Density of free oxide, ρ_f	Density of oxide built in the glass, ρ_g
1	Al_2O_3	3.99	2.50
2	B_2O_3	1.85	2.80
3	BaO	5.72	7.00
4	CaO	3.32	3.90
5	K_2O	2.32	2.80
6	Li_2O	2.43	2.70
7	MgO	3.65	3.30
8	Na_2O	2.27	3.10
9	P_2O_5	2.30	2.55
10	SiO_2	2.20	2.28
11	TiO_2	3.84	3.80
12	ZnO	5.49	5.60
13	ZrO_2	5.49	5.30

5.3.1 Factors for Calculating the Glass density

The simplest way to calculate the density is to use the direct application of the additives formula. We can calculate the density using the data of density for the oxides bound in the glass, as long as we know the weight percentage of the oxide and its density when it will be bound in the glass. The densities of a number of oxides widely used in medical glasses are shown in Table 5.1.

By following a number of simple steps it is possible to calculate the density for a glass that contains a number of oxides, A, B, and C with corresponding densities of oxides bound in the glass ρ_A, ρ_B, and ρ_C and weight percentages of X_A, X_B, and X_C.

$$\text{Volume of oxide A in the glass per unit mass, } V_A = \frac{\text{wt\% of oxide A}}{100\rho_A}$$

$$\text{Volume of oxide B in the glass per unit mass } V_B = \frac{\text{wt\% of oxide B}}{100\rho_B}$$

$$\text{Volume of oxide C in the glass per unit mass } V_C = \frac{\text{wt\%ofoxideC}}{100\rho_C}$$

$$\text{Calculated density: } \rho = m/V = \frac{1}{V_A + V_B + V_C + \cdots}$$

Examples of measured and calculated density values of Li_2O–Al_2O_3–SiO_2 glasses are shown in Table 5.2. It can be seen in each case that the measured densities of the

Table 5.2 Density values of Li_2O–Al_2O_3–SiO_2 glasses measured after Karapetyan et al. (1980) and calculated using the additives formula

Composition, wt%			The volume of each oxide in the glass per unit mass = wt% of oxide/100ρ_{oxide}			Total volume of oxides in the glass per unit mass $= V_{Li_2O} + V_{Al_2O_3} + V_{SiO_2}$	Calculated density $= 1/(V_{Li_2O} + V_{Al_2O_3} + V_{SiO_2})$	Density % difference	Measured ρ (g/cm^3)
Li_2O	Al_2O_3	SiO_2	Li_2O	Al_2O_3	SiO_2				
29	14	57	0.11	0.06	0.25	0.41	2.42	0.37	2.41
30	9	61	0.11	0.04	0.27	0.41	2.41	0.49	2.4
23	12	65	0.09	0.05	0.29	0.42	2.39	0.03	2.39
27	8	65	0.10	0.03	0.29	0.42	2.40	0.74	2.38
13	21	66	0.05	0.08	0.29	0.42	2.37	0.93	2.35
25	9	66	0.09	0.04	0.29	0.42	2.39	0.5	2.38

Table 5.3 The glass density difference factor for a range of commonly used oxides

Oxide	Factor for calculating density change (Y_m, g/cm³) for oxides after Huggins
SiO_2	−0.0024
Al_2O_3	+0.0018
CaO	+0.0106
MgO	+0.0050
BaO	+0.0173
Na_2O	+0.0050
K_2O	+0.0028
B_2O_3	−0.0036
ZnO	+0.0145

NB: The +ve or −ve sign indicates an increase or a decrease in density when adding the oxide

glasses are higher than the calculated densities based on the additives calculations. The glass is found to have higher densities than their corresponding crystalline oxides. The difference between the calculated and measured densities is less than 1% and in some glasses can reach up to 3%. There will always be a difference between the calculated and measured density of glass, which is possibly because of the crystalline nature of the oxide and the amorphous nature in the glass.

5.3.2 Calculation of Density Change with Compositional Variation

Adding or removing a fraction of any oxide in a silicate glass will change the glass density. Whenever one is dealing with silicate glasses it is useful to be able to determine how the change in the composition of a glass will affect the density of the glass without necessarily having to calculate the density for each glass. Elliott (1945) describes how to calculate the change in density of silicate glasses after an increase or decrease of an oxide in the glass batch.

The change in the glass density can be calculated from the following equation:

$$\Delta\rho = Y_1 \cdot \Delta X_1 + Y_2 \cdot \Delta X_2 + \cdots + Y_n \cdot \Delta X_n$$

Where $\Delta\rho$ is the change in density, ΔX the change in the component oxide in wt%, Y is the glass density difference factor (see Table 5.3).

However, conveniently the glass density difference factors for the calculation of the density change on increasing the content of an oxide in silicate glasses that contain 69–73% silica are presented in Table 5.3. The factors are applicable for glasses with density between 2.45 and 2.53 g/cm³ (Elliott, 1945).

Some examples of how to calculate the density change on addition of different oxides in the glass batch using Y_m factors are presented below.

Example 1

Replacement of 1 wt% silica in a silicate glass by 1 wt% alumina
For silica: $Y_1 \cdot \Delta X_1 = (-1.0) \cdot (-0.0024) = +0.0024$
For alumina: $Y_2 \cdot \Delta X_2 = (+1.0) \cdot (+0.0018) = +0.0018$
Density change $\Delta\rho = Y_1 \cdot \Delta X_1 + Y_2 \cdot \Delta X_2 = +0.0024 + (+0.0018) = +0.0042\,\mathrm{g/cm^3}$

Example 2

Replacement of 1 wt% CaO in a silicate glass by 1 wt% Na_2O
For CaO: $Y_1 \cdot \Delta X_1 = (-1.0) \cdot (+0.0106) = -0.0106$
For Na_2O: $Y_2 \cdot \Delta X_2 = (+1.0) \cdot (+0.0050) = +0.0050$
Density change $\Delta\rho = Y_1 \cdot \Delta X_1 + Y_2 \cdot \Delta X_2 = -0.0106 + (+0.0050) = -0.0056\,\mathrm{g/cm^3}$

5.4 Calculation of the Refractive Index of a Glass

The optical properties of glasses and glass ceramics are characterized by optical constants, which include the refractive index, absorptivity, and reflectance. The optical constants change with the radiation wavelength expressed in nanometers or angstroms. One of the important optical properties for medical glass applications is the refractive index, which can be computed from the chemical composition. Many glasses and ceramics are valued for use in dentistry particularly because of their optical translucency. Thus, the use of glasses or ceramics in dental restorations requires knowledge of the factors controlling the refractive index and its dependence on the wavelength of light.

5.4.1 Definition of the Refractive Index

The refractive index (or index of refraction) of a medium is a measure for how much the speed of light is reduced inside the medium. When a beam of monochromatic light passes through an air atmosphere (or ideally in vacuum) with a maximum speed W_1 into optically denser materials such as a glass, the beam of light slows down and it is deflected and refracted from its original direction as shown in Fig. 5.1. For example, a glass will typically have a refractive index of 1.5, which means that in the glass, light travels at 0.67 times the speed of light in a vacuum (1/1.5=0.67).

The extent of the beam refraction is represented by the refractive index. Thus, the refractive index is defined as the ratio of the speed of the monochromatic light having a wavelength λ in a vacuum to its speed in the glass. The refractive index is a ratio of two speeds and therefore is a dimensionless property.

$$\text{Refractive index}\,(\eta) = \sin\alpha_1 / \sin\alpha_2 = W_1 / W_2$$

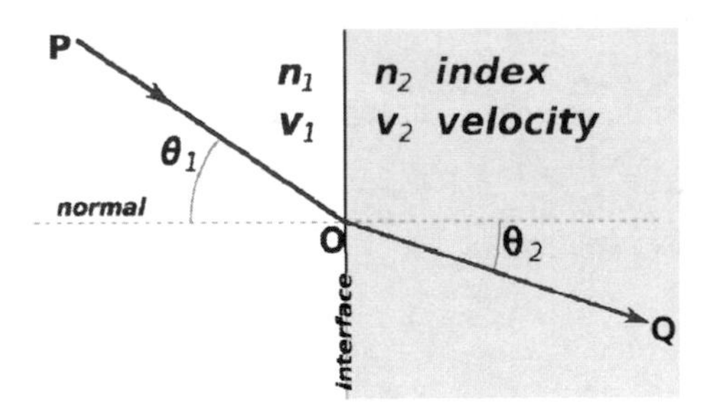

Fig. 5.1 Monochromatic light passing from air into dense materials

where α_1 is the angle of incidence, α_2 is the angle of refraction, W_1 is the maximum speed of the beam in air or vacuum, W_2 is the speed of the beam in the glass.

The refractive index will vary with the wavelength, so the wavelength at which it is measured must be specified. For convenience, the refractive index is related to a universally agreed set of wavelengths as follows:

C for 656.3 nm (red hydrogen line)
D for 689.3 nm (yellow sodium line)
E for 486.1 nm (blue hydrogen line)

Mostly, the refractive index is specified for the *D* line, as this lies approximately at the center of the visible spectrum. The difference between the refractive indices values at the blue hydrogen wavelength *E* (486.1 nm) and the red hydrogen wavelength *C* (656.3 nm) lines is used as a measure of the wavelength dependence of the refractive index and is called dispersion.

The refractive index of glasses affects their brilliance. The index of refraction has frequently been used to describe the opacity. All opacifiers have high indices of refractions, it might be concluded that a glass with a high index of refraction would be the most desirable for developing the opacity. The index of refraction tends to increase with the specific gravity, so a glass with a higher density will also have a higher refractive index.

5.4.2 Calculation of the Refractive Indices for Glasses with Known Density

The factors for the calculation of the refractive indices of glasses are shown in Table 5.4. The index of refraction for a glass with known density can be determined from the following equation:

$$\text{The index of refraction } (\eta_D) = 1 + \mathrm{d}\sum f_M \times r_{M \cdot D}$$

$$\text{Weight fraction of oxide } (f_M) = \frac{\text{oxide in wt\%}}{100}$$

Table 5.4 Factors ($r_{M\cdot D}$) required in the calculation of the refractive indices of glasses (after Huggins and Sun, 1945)

Oxide	Al_2O_3	B_2O_3	BaO	K_2O	Li_2O	MgO	Na_2O	SiO_2	P_2O_5	TiO_2	ZnO	CaO	ZrO_2
$r_{M\cdot D}$	0.207	0.215	0.2257	0.202	0.308	0.210	0.1941	0.20826	0.202	0.313	0.1499	0.227	0.209

Table 5.5 The indices of refraction and densities calculated according to the previously mentioned methods

Composition, wt%			Calculated density	Measured refractive index	Calculated refractive index
K_2O	MgO	SiO_2			
10	7	83	2.38	1.50	1.49
7	11	82	2.39	1.49	1.50
10	10	80	2.40	1.49	1.50
15	5	80	2.38	1.50	1.49
4	17	79	2.43	1.50	1.51
7	15	78	2.42	1.50	1.50
16	8	76	2.41	1.50	1.50
22.9	9.8	67.3	2.46	1.50	1.51
23	13	64	2.49	1.50	1.51
35	5	60	2.48	1.51	1.51
41.8	2.9	55.4	2.49	1.51	1.51

The measured values of refractive indices after Faick et al. (1935)

Table 5.6 Indices of refraction of lithium aluminum silicate glasses are calculated according to the method described above

Composition, wt%			Measured density, g/cm³	Weight fraction of oxide (f_M). Factor ($r_{M\cdot D}$)					Calculated refractive index
Li_2O	Al_2O_3	SiO_2		Li_2O	Al_2O_3	SiO_2	$\Sigma f_M \cdot r_{M\cdot D}$	$d\Sigma f_M \cdot r_{M\cdot D}$	
29	14	57	2.41	0.09	0.03	0.12	0.24	0.57	1.57
30	9	61	2.4	0.09	0.02	0.13	0.24	0.57	1.57
23	12	65	2.39	0.07	0.02	0.14	0.23	0.55	1.55
27	8	65	2.38	0.08	0.02	0.14	0.24	0.56	1.56
13	21	66	2.35	0.04	0.04	0.14	0.22	0.52	1.52
25	9	66	2.38	0.08	0.02	0.14	0.23	0.55	1.55

The calculated refractive indices for the glass compositions where the density is known are consistently closer to the measured values. Measured and calculated indices of refraction for a range of compositions of a potassium magnesium silicate and a lithium aluminum silicate glass are shown in Tables 5.5 and 5.6, respectively.

Indices of refractions and densities for a range of different phosphate glass compositions are shown in Table 5.7.

Table 5.7 Refractive indices and densities of different phosphate glasses calculated according to the above methods

Chemical composition, wt%					Calculated density, g/cm^3	Calculated refractive index
P_2O_5	CaO	MgO	Na_2O	TiO_2		
73.94	11.26	1.51	10.2	3.08	2.74	1.57
70.36	11.1	0.8	15.36	2.38	2.74	1.56
69.34	13.59	1.79	12.29	2.99	2.78	1.58
69.17	13.64	1.64	13.53	2.03	2.78	1.57
68.97	11.9	1.43	15.68	2.02	2.77	1.57
67.48	8.1	2.91	19.86	1.65	2.76	1.56
66.42	11.64	1.68	19.01	1.25	2.78	1.57
65.36	15.94	2.17	14.5	2.03	2.81	1.58
61.32	18.96	2.51	17.21	0	2.84	1.58
59.91	18.58	2.42	16.77	2.32	2.86	1.59
58.06	13.34	3.31	16.44	8.84	2.87	1.61
59.75	18.18	4.51	16.65	0.91	2.85	1.59
58.31	15.16	3.76	13.89	8.88	2.88	1.62
56.83	17.64	4.46	16.07	5	2.89	1.61

5.5 Estimation of the Coefficient of Thermal Expansion of a Glass

Whenever it is desirable to veneer or coat a substrate with a glass, it is not enough to take just any glass and fuse it to the substrate. I would be by some lucky chance if the glass layer were to match with the substrate to form a uniform closely covering coating. Much more likely is that on cooling, the glass will separate from the substrate either by delamination or fragmenting due to the differential cooling contraction. The thermal expansion on heating and the contraction on cooling are the most important parameters responsible for the adhesion of a glass or glass ceramic to a substrate. The thermal expansion and contraction of glass is controlled by the chemical composition and thermal history of the coating.

In general, the thermal expansion of a glass is related to the vibrations of the atoms in the material as a result of the change in the thermal energy. In solid glasses, the vibrations are restricted by the strong metal–oxygen bonds.

5.5.1 Definition of the Linear Coefficient of Thermal Expansion

The expansion of a glass is either specified by the linear thermal expansion or the volume expansion, although the former is much more commonly used. The linear thermal expansion coefficient (α) is defined as the relative elongation of a rod

(1 cm in length) as a result of an increase in temperature of 1 K. The temperature dependence of the length of a rod can be expressed by the following equation:

$$L_t = L_0(1 - \alpha\Delta t)$$

The coefficient of linear thermal expansion, α, is:

$$\alpha = (L_t - L_0)/L_0\,\Delta t$$
$$= \Delta L/L_0\,\Delta t$$

Where L_0 = original sample length at room temperature, L_t = the final sample length at a temperature t, Δt = Change in temperature, ΔL = Change in length = $L_t - L_0$.

The linear thermal expansion coefficient serves for a specified temperature range and sometimes called the mean linear thermal expansion coefficient. The commonly used abbreviation for the linear thermal expansion coefficient is TEC (thermal expansion coefficient).

5.5.2 Coating a Glass to a Metal Substrate

Whenever a glass is to be applied as a coating on a substructure, such as used in metal-ceramic restorations in dentistry, the glass layer must match the substrate to form a uniform, closely covering coating. When a glass powder is fused onto a comparatively inert metal substrate with a different coefficient of thermal expansion, differential contraction takes place on cooling. Stresses will develop between the metal substrate and the ceramic coating, which finally can become sufficiently large so as to cause delamination or fracture of the ceramic coating.

To eliminate these differential stresses as much as possible, it is of primary importance that the respective thermal expansions of the substrate and the glass shall coincide as closely as possible over the whole range from the lowest annealing temperature to room temperature. The chemical composition of the coatings will thus vary widely as these will depend on the type of substrate used.

In the case of a metal substructure, it is preferred that the TEC of the glass and the substrate are not exactly equal. It is usually arranged such that the substrate contraction shall be slightly higher than that of the glass coating. In this way, the glass will be exposed to a little compression, while putting the metal substrate under a little tension.

On coating the metal alloy with a glass, the glasses are found to be very sensitive to very small differences in their TEC. Some oxides for elements with a valancy ≥ 2, such as Al_2O_3, CaO, and BaO, are capable of raising the TEC in the part of the dilatometeric curve above the transition temperature of the glass T_g. So this part of the curve becomes steeper above T_g.

The value of the TEC can be altered by varying the glass content, especially with the use of Na_2O, K_2O, and NaF_2. Glasses with large amounts of these components

Table 5.8 Thermal expansion additivity factors after Mayer and Havas (1930)

Oxides	Factors × 10^{-7}/°C
SiO_2	0.8
Al_2O_3	5.0
B_2O_3	0.10
Na_2O	10.0
K_2O	8.5
ZnO	2.1
CaO	5.0
MgO	0.10
BaO	3.0
P_2O_5	2.0
SnO_2	2.0
TiO_2	4.1
ZrO_2	2.1
Na_3AlF_6	7.4
NaF	7.4
AlF_3	4.4
CaF_2	2.5
Cr_2O_3	5.1
NiO	4.0
MnO_2	2.2
CuO	2.2
CoO	4.4
Sb_2O_3	3.6
CeO_2	4.2
Li_2O	2.0
Fe_2O_3	4.0

are expected to display a high TEC. In contrast, glasses with a high content of the alkaline earth oxides MgO, Li_2O, ZnO, CaO have a lower TEC. Thus by increasing or reducing these oxides, it is possible to have very effective control over the coefficient of thermal expansion of the glass.

5.5.3 Estimation of the Thermal Expansion from Glass Chemical Composition

Various attempts have been made to determine a unified method for predicting the TEC from the chemical composition of the glass. Approximate additivity factors have been identified, which can be used for the calculation of the volumetric thermal expansion coefficient of the glass (Table 5.8).

The product of the additivity factor for each oxide multiplied by its weight percentage in the glass is calculated and the total of the products for all the oxides in the glass represents the volumetric thermal expansion coefficient [product = Σ(factor × oxide in wt%)].

Table 5.9 Examples of how to calculate the TEC from the chemical composition

Oxide	wt%	Factor $\times 10^{-7}$/°C	CTE $\times 10^{-7}$/°C
K_2O	11.00	8.50	93.50
Na_2O	4.20	10.00	42.00
Al_2O_3	13.60	5.00	68.00
SiO_2	61.00	0.80	48.80
CaO	1.70	5.00	8.50
B_2O_3	2.50	0.10	0.25
TiO_2	6.00	4.10	24.60
Calculated cubic TEC	285.65		
Calculated linear TEC	95.22		
Measured TEC	103.4		

The cubic thermal expansion or the volumetric thermal expansion of isotropic materials (example; glass) approximately equals 3 times as the linear thermal expansion

Table 5.10 Calculation of the approximate TEC of a bioglass

	Oxide composition					
Glasses	SiO_2	Na_2O	K_2O	CaO	MgO	P_2O_5
Bioglass (wt%)	45.0	24.5	0.0	24.5	0.0	6.0
Oxide additivity factor	0.8	10.0	8.5	5.0	0.45	2.0
Oxide wt% $\times$ factor	36.0	245	0.0	122.5	0.0	12.0
Volumetric thermal expansion coefficient = total of products						415.5
Calculated Linear TEC = total/3 $\times 10^{-7}$/°C						138.5
Measured TEC (10^{-7}/°C)						151

The linear thermal expansion coefficient can be calculated as follows:

$$\alpha = \frac{a\rho_1 + b\rho_2 + c\rho_3 + d\rho_4 + \cdots}{3},$$

where α = thermal expansion coefficient; a, b, c = additive factors for each oxides; ρ_1, ρ_2, ρ_3, ρ_4 = weight percentage of each oxide.

Examples of how to calculate the TEC from the chemical composition are shown in Table 5.9 and an example of calculation of the approximate TEC of a bioglass is shown in Table 5.10.

As shown by the values of thermal expansion, the calculated value is somewhat different from the measured value as the values are affected by many factors including the chemical composition of the glass that may have affected by the volatilization of low melting constituting oxides during melting as P_2O_5 and B_2O_5, in addition to the expected volatilization of alkali oxides as Na_2O at higher temperature.

The change of the expansion coefficient with the change in the chemical composition is shown in the family of glass in the Li_2O–Al_2O_3–SiO_2 system as shown in Table 5.11. The coefficient of thermal expansion of each composition was calculated.

Measured and calculated thermal expansion of different glass compositions are also shown in Table 5.12.

Table 5.11 TEC of some glasses in the system Li_2O–Al_2O_3–SiO_2 glasses

Glass	Composition, wt%			Measured $\alpha \times 10^{-7}$/°C (30–500°C)	Calculated 10^{-7}/°C
	Li_2O	Al_2O_3	SiO_2		
Li_2O–Al_2O_3–$4SiO_2$	8	27.4	64.6	66.6	68.2
Li_2O–Al_2O_3–$6SiO_2$	6.1	20.7	73.2	55.5	58.1

Table 5.12 Measured and calculated thermal expansion of different glass compositions

Oxides	Glasses			
	Glass 1	Glass 2	Glass 3	Glass 4
Na_2O	16.72	15.82	14.95	13.70
K_2O	3.20	4.18	5.13	3.00
CaO	0.09	0.12	0.14	0.09
Al_2O_3	5.28	6.88	8.44	4.97
B_2O_3	20.00	18.02	10.12	14.24
SiO_2	44.05	44.52	45.07	54.00
CaF_2	8.00	7.84	7.70	7.53
CoO	0.67	0.65	0.84	0.63
MnO_2	2.00	1.96	2.92	1.88
Measured α (10^{-7}/°C)	95.0	96.0	99.3	82.0
Calculated α (10^{-7}/°C)	95.3	97.6	100.8	86.1

Further Reading

Andrews, A.I., How, E.E.: The effect of fluorides on the properties of white sheet-iron enamels. J. Am. Ceram. Soc. **17**(1–12), 288–291 (1934)

Andrews, A.I., Smith, R.K.: The thermal expansion of sheet-iron ground-coat enamels. J. Am. Ceram. Soc. **16**(1–12), 328–337 (1933)

Brauer, D.S., Rüssel, C., Kraft, J.: Solubility of glasses in the system P_2O_5–CaO–MgO–Na_2O–TiO_2: experimental and modeling using artificial neural networks. J. Non-Cryst Solids **353**, 263–270 (2007)

Elliott, R.M.: Glass composition and density changes. J Am Ceram Soc **28**(11), 303–305 (1945)

Faick, C.A., Young, J.C., Hubbard, D., Finn, A.N.: Index of refraction, density and thermal expansion of some soda-alumina-silica glasses. Bur. Stand. J. Res. **14**, 133 (1935)

Geoffrey, C., Maitland, M.A., Gordon, G.S.: Ceramists Handbook Stoke-On Trent, p. ST1 3PG. Podmore, Shelton (1964)

Hall, F.P.: The influence of chemical composition on the physical properties of glazes. J. Am. Ceram. Soc. **13**(3), 182–199 (1930)

Huggins, M.L., Sun, K.H.: Calculation of density and optical constants of a glass from its composition in weight percentage. J. Am. Ceram. Soc. **26**(1), 4–11 (1945)

Lopez-Esteban, S., Saiz, E., Fujino, S., Ku, T.O., Suganuma, K., Tomsia, A.P.: Bioactive glass coatings for orthopedic metallic implants. J. Eur Ceram Soc **23**, 2921–2930 (2003)

Mayer, M., Havas, B.: The coefficient of thermal expansion of sheet, iron enamels. Sprechsaal **42**, 497–9 (1909a)

Mayer, M., Havas, B.: The function of fluorine compounds in enamels. Sprechsaal **42**, 460–61 (1909b)

Mayer, M., Havas, B.: Expansion coefficient of sheet iron enamels. Chew-Ztg. **33**, 1314 (1914)
Roedder, E.W.: The system K_2O-MgO-SiO_2; Part 1. Am. J. Sci. **249**, 81 (1951)
Tick, P.A.: Zirconium-alkali fluorophosphate glasses. Phys Chem Glasses **23**(5), 73–76 (1982)
Volf, M.B.: Mathematical Approach to Glass, Glass Science and Technology. Elsevier, Amsterdam (1988)

Chapter 6
Design and Raw Materials of Medical Glasses

This chapter emphasizes the glass structure and conditions of glass formation. In addition, the chapter gives a brief view on how to optimize the glass properties through mastering the chemical composition via mathematical calculations.

6.1 Design of Glass Composition

Glass is an inorganic material made by fusion and is able to be cooled without being crystallized. In fact, this definition in the current times becomes too short as there are other glasses that can be prepared using chemical techniques that do not involve cooling such as sol–gel. Although almost any element in the periodic table could be present in the composition of glass, few compositions would be found that contain no silicon, boron, or phosphorous in considerable quantities. Such elements and their corresponding oxides are simply referred to as glass formers. Another group of elements that approach glass forming properties, but do not quite meet the requirements are called Intermediates (or in oxide form called Intermediate oxides). Another group of elements that practically lack the ability to form a glass are called Modifiers (or in oxide form called Modifying oxides). So, almost any glass can be described in terms of the relative amounts of glass formers, intermediates, and glass modifiers. Furthermore, other oxides called glass stabilizers (as CaO, MgO, BaO) impart the glass a very high degree of chemical resistance and control the forming operation of glass. A functional classification of the common glass oxide categories is shown in Table 6.1.

E. El-Meliegy and R. van Noort, *Glasses and Glass Ceramics for Medical Applications*,
DOI 10.1007/978-1-4614-1228-1_6,

Table 6.1 Functional classification of some oxides used in manufacturing glass

Network formers	Intermediates	Network modifiers
B_2O_3	Al_2O_3	MgO
SiO_2	Sb_2O_3	Li_2O
GeO_2	ZrO_2	BaO
P_2O_5	TiO_2	CaO
V_2O_5	PbO	SrO
As_2O_3	BeO	Na_2O
	ZnO	K_2O

6.2 Raw Materials for Glass

The most important aspect in the selection of the raw materials for the glass industry is to take into your consideration the purity. Also, the composition of the raw materials must be constant within fairly close limits, so proper control over the glass ceramics can be achieved.

A general glass batch for commercial use usually consists of 5–12 components. Most of the glass batches are made up of four to six components, mainly silica, alumina in the form of feldspar, limestone, albite, and/or nepheline. The remainder is composed of several additives of up to 20 materials as minor ingredients. These latter additions are added to perform certain functions in the glass. The primary purposes are as the following:

1. Glass fining components such as 0.1–0.3% as sodium chloride
2. Glass decoloration provided by rare earth oxides
3. Glass color development with the addition of vanadium oxide
4. Oxidizing agents
5. Melting aids such as fluorspar
6. Opacifiers, that can be added by up to 20%
7. Nucleating agents such as TiO_2, niobium oxide

As this book is concerned with medical/dental ceramics that will be used in the oral cavity and as implants, the materials selected to be used must be of high purity, free of coloring impurities, and subjected to different cycles of purification.

The starting precursor of the anticipated dental ceramic is either a bulk glass or a very fine glass powder. So it is very important to throw light on what might be suitable raw materials for the preparation of glass ceramics for the dental ceramics industry.

The proposed dental glass must be of a very high purity, made from either pure natural minerals or pure ceramic oxides, free of discoloring oxides or easily soluble toxic oxides. So the glass prepared for the production of dental materials requires special purity starting materials compared with traditional glasses for other applications.

The natural minerals to be used must pass through a strict dressing and purification process to reach very high purity. Some impurities are compatible with dental

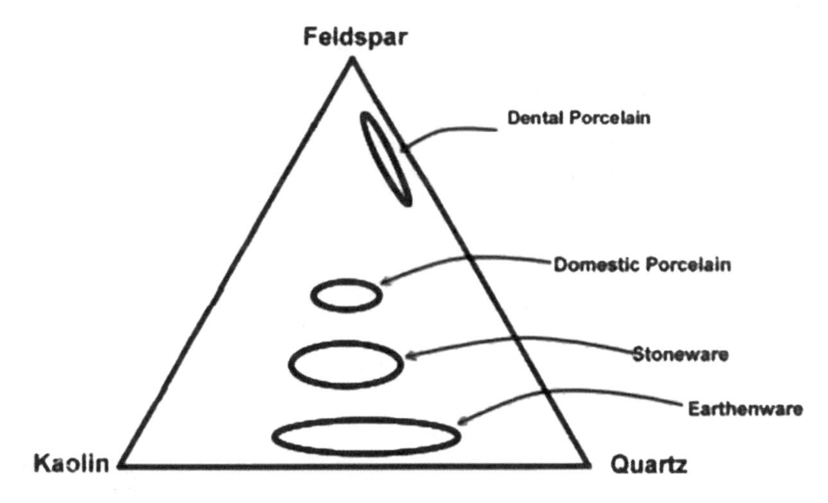

Fig. 6.1 Triaxial diagram of traditional porcelain

glasses such as Na_2O, P_2O_3, MgO, and CaO in potash feldspar, so their presence does not represent great problems. On the other hand, other oxides such as iron oxides will result in a severe problem of color disturbance, especially during the process of color development.

The most important raw materials for the manufacture of traditional porcelain are kaolin, quartz, and feldspar and their derivatives. The compositions of the different types of porcelains are shown in the triaxial diagram in Fig. 6.1.

Feldspar, nepheline, quartz, volcanic glass, and other raw materials supporting sintering and melting can be used in glass preparation in a commercial approach to the formation of a glass. On the other hand, these natural minerals must be used within suggested criteria controlling their chemical analyses. Since the consumption of the natural minerals in dental glass is limited, compared to their consumption in traditional ceramics and glasses, the choice of high purity materials is easier.

6.2.1 Feldspars Fluxing Agent

A fluxing material lowers the temperature of fusion. The flux material has a high content of alkali and alkaline earth elements. However, the alkali content of the flux is not the only factor that controls their function. The flux material must be free of iron oxide to avoid the discoloration of the products. The particle size of the flux is also of vital importance. The reactivity and effectiveness of the flux are controlled by the fineness. The fineness results in an increase in the surface area and this increases the solubility of alkalis.

Feldspars; orthoclase ($K_2O{\cdot}Al_2O_3{\cdot}6SiO_2$), albite ($Na_2O{\cdot}Al_2O_3{\cdot}6SiO_2$), and anorthite ($CaO{\cdot}Al_2O_3{\cdot}2SiO_2$) are a large group of rock forming aluminosilicate minerals and an excellent source of potassium, sodium, and calcium. Only alkali containing feldspars, potassium and sodium feldspars are used in the glass industry. Potash feldspar or orthoclase ($KAlSi_3O_8$) and soda feldspar or albite feldspar ($NaAlSi_3O_8$) are the main industrial ceramic fluxes. In order to be useful as a component in dental ceramics, these feldspathic minerals must be free of impurities like undesirable quantities of transition metal cations such as iron, chromium, nickel, and manganese. In addition, the feldspar content should be as high as possible in the raw materials.

Potassium feldspar gives the product a higher viscosity, higher strength, and greater transparency compared with sodium feldspar. The grain size of the milled feldspar has a substantial influence on the fusion and transparency. The finely milled feldspar vitrifies and fuses at a lower temperature. The transparency of porcelain is greatly improved with the use of feldspar and quartz with a grain size in the range of 10–50 μm. The choice of feldspar suitable for dental ceramics is facilitated through investigation by a heating microscope, which allows for the determination of the shrinkage, softening and fusion temperatures and helps to examine the color.

Typical feldspar should have the following characteristics.

1. Start of shrinkage at 950°C
2. Start of softening at 1,100°C
3. Start of rounding corners at 1,270°C
4. Fusion Semi sphere point at 1,550°C

6.2.2 Silica (SiO_2)

Silica acts as the backbone of dental ceramics. The ceramic bodies contain two types of silica either combined silica (in the form of silicates taken from feldspar and talc) or free silica (in the form of quartz, sand, and flint). The free silica has a much greater thermal expansion than combined silica.

Three basic silica crystalline forms are found, which includes quartz, tridymite, and cristobalite. The largest amounts of silica are found in the reacted form in aluminosilicates. Pure silica is obtained generally from quartz sands. Silica frequently occurs in a glassy fused form, which is preferable in any ceramic as fused silica has a very low thermal expansion coefficient, it being less than one tenth that of other crystalline forms of silica. Therefore, fused glassy silica has a high thermal shock resistance compared with crystalline silica.

Phase transformation in silica is a very special phenomenon, which affects the thermal properties in ceramics. Silica exists via inversions and conversions in different forms. There are three crystalline forms of silica namely quartz, tridymite, and cristobalite. These three forms have the same chemical formula (SiO_2), but differ in their crystal structures. Silicon and oxygen atoms are less closely packed in tridymite and cristobalite than in quartz. Hence, tridymite and cristobalite show a lower specific gravity of 2.32 and 2.28, respectively when compared to 2.65 for quartz Roy and Roy (1964).

Cristobalite and tridymite are the higher temperature phases. Quartz is the stable phase form up to 870°C. Tridymite is the stable phase between 870 and 1,470°C, while cristobalite is stable phase between 1,470 and 1,710°C. Above 1,710°C, silica is exposed to fusion resulting in the glassy fused silica form. The changes from one crystalline form to another form are very slow and reversible. The actual formation of tridymite from quartz or cristobalite may not take place at all without the presence of mineralizers. Addition of one of the fluxing agents as Na^{+}, Li^{+}, or Ca^{+2} combined with Al^{+3} speed the inversion temperature of cristobalite into tridymite and the lowers the temperature of inversion.

6.2.3 *Alumina (Al_2O_3)*

Alumina (Al_2O_3) is used alongside silica to become the second most widely utilized glass raw material. Alumina is produced from bauxites (in the form of aluminum hydroxides). Alumina is available in powder form with a wide range of sizes, ranging from fine (submicron) sinterable powders, to mm size grinding grits, or in fibrous or single crystal form. Alumina is considered a main component of dental glasses. It is chemically stable against most environments except hydrofluoric acid and some molten salts. Alumina also helps to prevent devitrification and adds strength to the glass products. Potash feldspar ($K_2O{\cdot}Al_2O_3{\cdot}6SiO_2$), albite ($Na_2O{\cdot}Al_2O_3{\cdot}6SiO_2$), anorthite ($CaO{\cdot}Al_2O_3{\cdot}2SiO_2$), and nepheline are the most common sources of alumina in glasses.

6.2.4 *Other Glass Raw Materials*

A wide range of other materials can be used in dental glasses and some may be present as impurities. A very small amounts of these is not a problem as long as they are free of discoloring oxides.

1. Barium aluminate ($3BaO{\cdot}Al_2O_3$) used in glass batches as a source of BaO.
2. Barium carbonate ($BaCO_3$) acts as a flux, having the lowest melting point of all alkaline earths. It helps to improve luster and imparts better translucency in glass ceramics.
3. Borax ($Na_2B_4O_7$): It is a water soluble, low temperature flux, which lowers the fusion point and promotes a smooth melt.
4. Calcium carbonate ($CaCO_3$): This is a common source of calcium and can improve the durability and hardness.
5. Calcium zirconium silicate ($CaO{\cdot}MgO{\cdot}SiO_2{\cdot}ZrO_2$): It acts as an opacifier in low temperature glasses.
6. Cerium oxide (CeO_2): It is used as an opacifier for special effects in dental glass ceramics.

7. Chrome oxide-green (Cr_2O_3): A versatile colorant used in glasses to produce various green tints.
8. Cobalt oxide (Co_3O_4): A colorant used in glasses to produce various blue tones.
9. Cobalt Oxide (Co_2O_3, CoO): Coloring agent in glass providing blue color.
10. Cryolite (Na_3AlF_6): This is a powerful flux with a relatively low melting point. It strongly reacts with SiO_2, Al_2O_3, and CaO to dissolve them in the melt.
11. Dolomite ($MgCa\ (CO_3)_2$): Useful as a source of calcium and magnesium and used as a high temperature flux and increases the chemical resistance of glasses.
12. Fluorspar (CaF_2): Used as an opacifier and flux for mica glasses and reduces the coefficient of thermal expansion.
13. Hydroxyapatite ($Ca(PO_4)_3OH$): Complex calcium phosphate bioceramics are used as substitutes for bone and, under the right conditions, hydroxycarbonate-apatite (HCA) develops as a thin surface layer upon exposure to body fluids.
14. Lithium carbonate (Li_2CO_3): Used as a flux in leucite glasses to decrease the fusion temperature and the thermal expansion coefficient without impairing the crystallinity.
15. Lithium fluoride: It is a powerful flux similar but stronger than the lithium carbonate.
16. Lithium oxide: Flux for glasses with low thermal expansion coefficient.
17. Magnesium carbonate ($MgCO_3$): Common source of magnesium in glasses. Imparts strength and decrease the shrinkage.
18. Magnesium oxide: Important high temperature flux in bioactive ceramics and glasses preparation and lowers the coefficient of thermal expansion.
19. Magnesium Phosphate ($Mg_2P_2O_8$): Act as an opacifier and can replace tin oxide in glasses and glazes for esthetic value. Works well both in low-alkali and high-alkali glasses and glazes.
20. Nepheline ($K_2O{\cdot}3Na_2O{\cdot}4Al_2O_3{\cdot}9SiO_2$): Higher in alkali and alumina, lower in silica than feldspars.
21. Potassium carbonate (K_2CO_3): It is a strong flux and can be used as a color modifier in glazes.
22. Soda ash (Na_2CO_3): Sodium carbonate is an active flux and also serves an important function as a deflocculant in preparing a ceramic slip. It increases strength and workability and reduces shrinkage and raises the thermal expansion coefficient.
23. Sodium phosphate ($Na_4P_2O_7$): Used as a flux and a source for sodium and phosphorous.
24. Spodumene ($LiAlSi_2O_6$): Replaces feldspar as a flux and also reduces the vitrification temperature.
25. Talc ($3MgO{\cdot}SiO_2{\cdot}H_2O$): Soft, platy hydrous magnesium silicate. Used to synthesize cordierite glass ceramics, a silicate of very low coefficient of thermal expansion.
26. Tin oxide (SnO_2): Effective opacifier to produce evenly opaque glass ceramics.
27. Titanium oxide: The strongest white pigment and opacifier for glasses.

28. Tri-calcium phosphate ($Ca_3(PO_4)_2$): A white amorphous powder that is insoluble in cold water but decomposes in hot water.
29. Tri-sodium phosphate ($Na_3(PO_4)_2$): It is a source of P_2O_5 and Na_2O.
30. Vanadium pentoxide (V_2O_5): It is a common glass colorant.
31. Wollastonite ($CaSiO_3$): This is used to reduce shrinkage and make brighter and smoother glazes and good fluxing.
32. Zinc oxide (ZnO): A useful, high temperature flux, increases the maturing temperature and produces bright, glossy colors.
33. Zinc zirconium silicate ($3SiO_2{\cdot}ZnO{\cdot}ZrO_2$): An opacifier, which gives brilliance to the color of a glaze. Usually combined with zircon compounds.
34. Zircon ($ZrSiO_4$): An opacifier which controls texture and craze resistance in glazes.
35. Zirconium spinel ($ZrSiO_4$): A synthetic zirconium silicate with excellent color stabilizing abilities.
36. Zirconium oxide (ZrO_2): Used as a nucleating agent and opacifier.

6.3 Melting of Glass Batches

Glasses are made by heating the batch (a mixture of calculated raw materials) at a sufficiently high temperature to assure the reaction of the materials through homogeneous melting. Raw materials are accurately weighed and after complete mixing are charged into a furnace to be melted between 1,200 and 1,600°C depending on the batch composition. The high temperature melting assures that vigorous reactions take place. Alkali carbonate reacts with silica and the other alkaline earth carbonates are dissociated, resulting in the evolution of volatiles such as carbon dioxide. Violent agitation of the molten mass is needed to ensure the evolution of gases. Eventually, complete fusion of the materials is affected and there remains the process of refining and removing bubbles from the melt. The rate of bubbles elimination is proportional to the square of its diameter, so that large bubbles rise to the surface and escape much more rapidly than fine ones. For this reasons, refining agents are included in the glass batch to aid in the evolution of gas bubbles at the later stages during melting process. Thus, refining agents help to produce large bubbles at the later stages during melting process and thus render the melt clear of large and fine bubbles.

6.4 The Glass Structure and Conditions of Glass Formation

Although the majority of natural manufactured solids are crystalline in nature, the amorphous materials are of great importance for both traditional and newly developed ceramics. The properties of the liquid silicates are essential parameters in the formulation of glasses, glazes, enamels, and glass ceramics. Solid glasses, of which

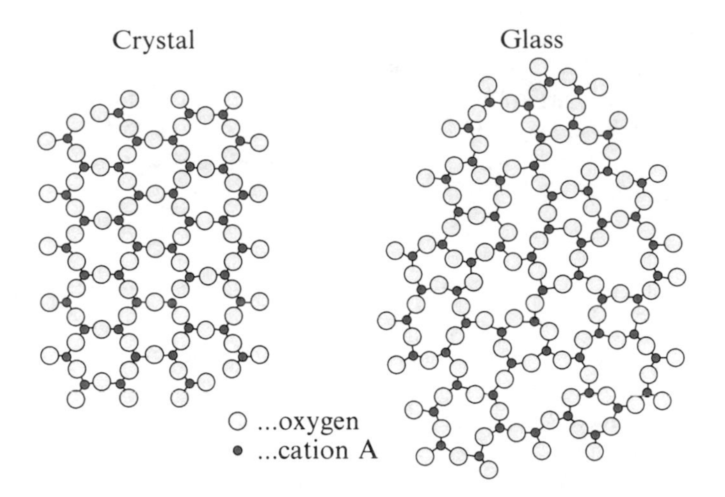

Fig. 6.2 The structure of a crystalline ceramics and an amorphous glass

the silicates are the technologically most important group, usually have a more complex structure than the liquids from which they are derived. In this part, we focus attention on the glass structure and microstructural features.

Glasses are usually formed by solidification from the melt. The structure of glasses can be clearly distinguished from that of liquids, since the glass structure is effectively independent of temperature. On cooling of the liquid, there is a discontinuous change in volume at the melting point if the liquid crystallizes.

However, if no crystallization occurs, the volume of the liquid decreases at about the same rate as above the melting point until there is a decrease in the expansion coefficient at a temperature range called the transition range. Below this temperature range the glass structure does not relax at the cooling rate. The expansion coefficient for the glassy state is usually about the same as that for the crystalline solid.

If slower cooling rates are used, the time needed for the structure to relax increases; the supercooled liquid persists to a lower temperature and the result is a higher density glass. Similarly, by heating the glassy material in the annealing range, in which slow relaxation can occur, the glass structure in time approaches an equilibrium density corresponding to the supercooled liquid at this temperature. The structures of the amorphous glass and the crystalline glass ceramics are shown in Fig. 6.2.

A concept useful in discussing the properties of glasses is the glass transition temperature, T_g, which corresponds to the temperature of the intersection between the curve for the glassy state and that for the supercooled liquid. Different cooling rates, corresponding to different relaxation times, give rise to a different configuration in the glassy state equivalent to different points along the curve of the supercooled liquid. In the transition range, the time for structural rearrangements is similar in magnitude to that of experimental observations. Consequently, the change in configuration toward the equilibrium structure of the glass in this temperature range occurs slowly with time. At somewhat higher temperatures, the structure corresponding to equilibrium at any temperature is achieved very rapidly. At substantially lower temperature the arrangement of the glass remains logically stable over long periods of time.

As the liquid is cooled from a high temperature without reaching the crystallizing temperature a bend appears in the volume–temperature relation. In this region, the viscosity of the melt increased to a sufficiently high value, typically 10^{-12} to 10^{-13} poises. The glass transition temperature increases with increasing cooling rate, as does the specific volume of a glass that is being formed. The maximum difference in specific volume that can be obtained with variations in the cooling rate is typically in the range of a few percent.

The XRD and electron diffraction patterns of glasses show that the glasses lack long range order of the constituting atoms. Glasses generally exhibit broad peak in place of the strong peaks that are also seen in the diffraction patterns of the corresponding crystals. This observation led to the suggestion that glasses are composed of assemblages of very small crystals. These minute crystals are termed crystallites with particle sizes smaller than about 0.1 μm and their presence results in a broadening of the XRD diffraction peaks. The broadening increases linearly with decreasing particle size. This model has been applied to both single and multicomponent glasses.

On the other hand, the random network model was originally proposed to explain glass formation as resulting from the similarity of structure and internal energy between crystalline and glassy oxides. The glasses are three-dimensional networks, lacking symmetry and periodicity. Four rules were suggested to control the formation of an oxide glasses.

1. Each oxygen ion should be linked to not more than two cations.
2. The coordination number of oxygen about the central cation must bc 4 or lcss.
3. Oxygen polyhedra share corners, not edges or faces.
4. At least three corners of each polyhedron should be shared.

The glass forming oxygen polyhedra are triangles and tetrahedra and cations forming such coordination polyhedra have been termed network formers. Alkali silicates form glasses easily where the alkali ions are supposed to occupy random positions distributed through the structure and provide local charge neutrality. The major function of alkali ions is viewed as providing additional oxygen ions that modify the network structure.

The alkali ions are called network modifiers. Cations of higher valence and lower coordination number than the alkali ions and alkaline earths may contribute in part to the network structure and are referred to as intermediates. In a general, the role of cations depends on the valence and coordination number and the relative value of the single bond strength.

Now there is a belief that crystallization during cooling is kinetically prevented and the structural elements in silicate glasses are randomly arranged and no unit of the structure is repeated at regular intervals. So, a necessary condition for an oxide to form a glass under a given quenching conditions is that the cation oxygen bond must be capable of some flexibility in order to permit a disordered structure and must also be strong enough to maintain the disorder when it is attained.

In order for an oxide to form a glass, the disordered state achieved by melting must be preserved to a sufficient degree on cooling. Under these conditions, the coordination number of the cation with respect to oxygen will be as low as 3 or 4.

In silicate glasses, the addition of alkali or alkaline earth oxides to silica increases the ratio of oxygen to silicon to a value greater than 2. The addition of alkali oxides as Li_2O, Na_2O, and K_2O or alkaline earth oxides as CaO, MgO, and BaO, breaks up the three-dimensional network with the formation of single bonded oxygen anions that do not participate in the network. For reasons of local charge neutrality, the modifying cations are located in the vicinity of single bonded oxygen. With divalent cations, two single bonded oxygen anions are required for each cation, for monovalent alkali ions, only one of such oxygen is required.

6.5 Glass Shaping into Block as Glass Ceramic Precursors

Whenever possible, glass ceramics are first shaped while molding in the glassy state. During shaping of the glass, internal stresses are produced due to the presence of temperature gradients within the glass during cooling. These stresses must be removed by exposing the glass to the proper annealing temperature or the stresses may result in fracturing of the shaped glass during reheat treatment through the crystallization process.

During heating, the composition of the glass will be progressively changed when viscous crystals are precipitated. In many cases, the effect of crystallinity is expected to increase the refractoriness of the residual glassy phase. In addition to the undesirable residual stresses, the effect of high rate of heating may result in glass deformation. Also rapid heating must be avoided, since it may result in excessive cracking of the glass ceramic. The danger arises because some of the crystals that are formed have relatively different densities from that of the glassy phase. This process of crystallization produces a volume change, which results in the generation of stresses in both the glass phase and the crystal phases.

With slow heating, these stresses can be relieved by viscous flow of the glassy phase. The slow increase of temperature should be continued until the upper crystallization temperature is reached. At the crystallization temperature, the crystallization will proceed rapidly without leading to glass ceramic deformation due to softening of the residual glass phase or remelting of the developed refractory crystalline phase. By maintaining the glass ceramic at the upper crystallization temperature for a suitable period, almost complete crystallization can be achieved such that eventually only a very small proportion of the residual glass phase will be present.

6.6 Transformation Range of Glass and Annealing of Glass Blocks

Glasses may be characterized by certain well-defined properties which are common to all, but different from those of liquids and crystalline solids. Glasses do not have a sharp melting point and cannot be cut, like crystals in a preferred direction. Glasses can flow like liquids under very high shear stress and show elasticity similar to crystalline solids.

Crystallization at or a little below the freezing point can occur rapidly on cooling the glass melt. There are, however, numerous glass melts, which are so viscous to hinder the easy movement of the atoms on cooling and in role reduce the rate of crystallization below the solidification temperature. Thus, if the crystallization rate is low enough, it is possible to go on cooling without crystallization.

As the melt cools, the viscosity continues to increase and below the freezing point it is called a supercooled liquid. The various pathways that a glass melt can take as it cools is illustrated in the volume–temperature diagram in Fig. 6.3. On cooling a liquid from the initial melt state A, the volume will decrease along AB. If the rate of cooling is slow and nuclei are formed, crystallization will happen at the freezing point, T_f. The volume will decrease rapidly from B to C, and the crystalline solid will contract along CD.

If the rate of cooling is sufficiently rapid, the liquid will supercool along BE which represent a continuation of AB to form a glass. At a certain temperature T_g (transition temperature) the graph undergoes a significant change in gradient and the glass continues contracting at a rate that is very similar to the contraction of the crystalline form. The transition temperature will decrease as the cooling rate is slowed. The rate of volume change decreases on holding the glass at temperature *T*, and the structure approaches an equilibrium state that corresponds to the lowest free energy.

Therefore, the transition point is defined as the temperature at which a molten mass is converted into an amorphous solid and is estimated to lie close to the

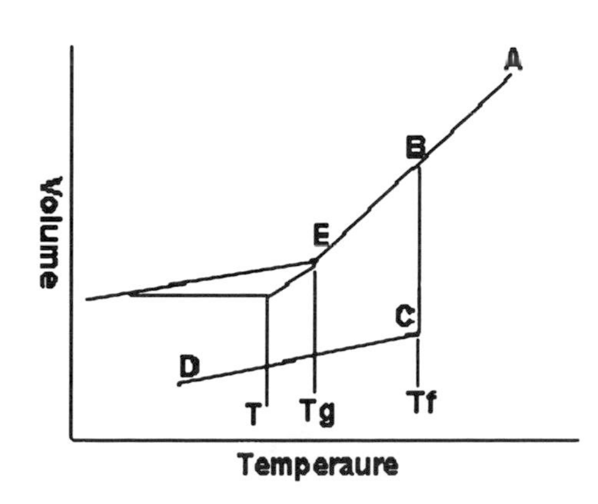

Fig. 6.3 The volume temperature diagram of glasses

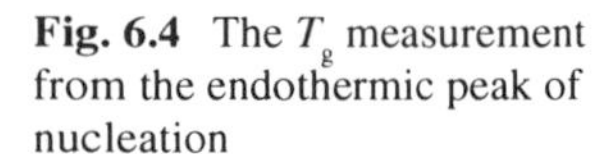

Fig. 6.4 The T_g measurement from the endothermic peak of nucleation

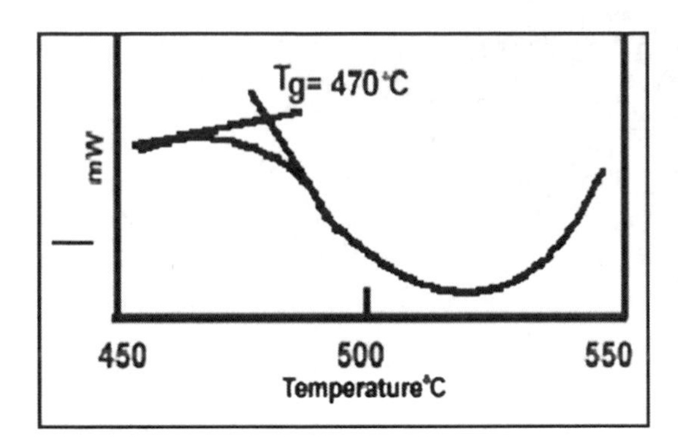

annealing point of a glass. The transition range has been defined as that range between the temperature at which a liquid melt is supposed to have been converted into an amorphous solid and that temperature commonly being held close to the annealing point of a glass. The transition temperature T_g can be calculated from a DTA curve of a glass as shown in Fig. 6.4.

Reference

Roy, D.M., Roy, R.: Tridymite-cristobalite relations and stable solid solutions. Am. Mineral. **49**, 952–961 (1964)

Part III
Manufacturing of Medical Glass Ceramics

Chapter 7
Design of Medical Glass-Ceramics

7.1 Glass Ceramic Fabrication

The art of glass-ceramics was first established with the publication of US Pat. No. 2,920,971. Classically, a glass-ceramic is made through controlled heat treatment of a precursor glass. The manufacture of a glass-ceramic involves three general steps:

- Preparation of a glass-forming batch from raw materials, oxides, and nucleating agents.
- The batch is melted and then cooled to a temperature within the transition range to yield a glass.
- The glass is then heated to a temperature above the transition range to perform glass crystallization.

A homogeneous glass batch is prepared using high-quality raw materials that are fused in a high temperature resistant crucible to form a homogeneous molten glass. The melt is then cooled into a glass body with the desired shape. The glass is then annealed at a temperature close to the transition temperature.

The crystallization process, the development of crystalline phases in the glass matrix, comprises two main steps. In the first step, the glass is heated to a temperature slightly above the transformation range and maintained for a sufficient time to achieve substantial nucleation. In the second step, the nucleated body is heated to a higher temperature, frequently above the softening point of the precursor glass to allow the growth of crystals on these nuclei. A homogeneously crystallized glass ceramic with relatively uniform size can be prepared with careful heat treatment.

The glass-ceramic is generally highly crystalline and the crystals themselves are very finely grained and preferably dispersed uniformly in the glass matrix. A wide range of microstructures can be created, including uniform crystal phases, interlocking crystals, and crystals with a wide variety of shapes and sizes.

E. El-Meliegy and R. van Noort, *Glasses and Glass Ceramics for Medical Applications*, DOI 10.1007/978-1-4614-1228-1_7,

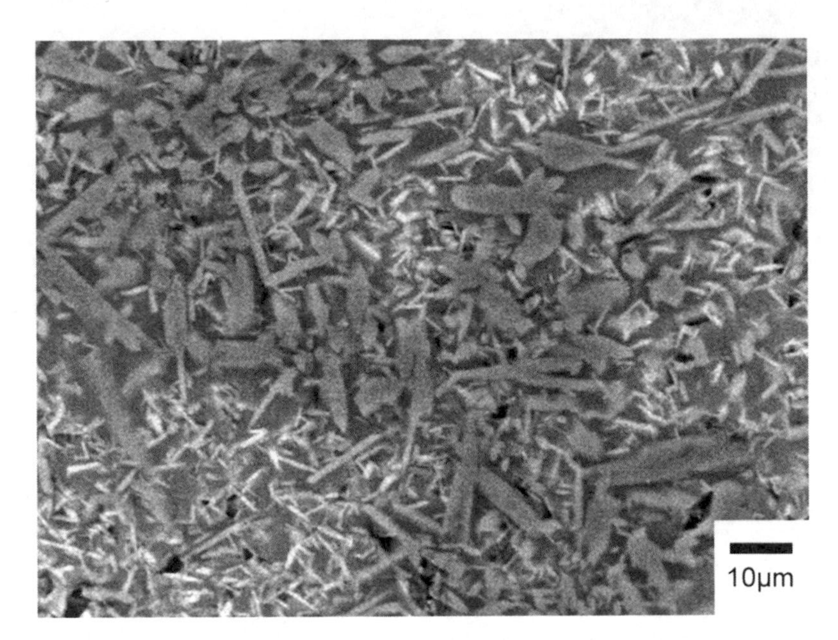

Fig. 7.1 SEM showing interlocking crystals of fluorophlogopite coupled with fine-grained β-spodumene

A good example is the interlocking of the fluorophlogopite crystals coupled with a relatively high percentage of fine-grained β-spodumene solid solution in Fig. 7.1. The microstructure promotes the highest mechanical strength and reduces the expansion coefficient to less than $35\times10^{-7}°C^{-1}$ compared with an expansion coefficient for fluorophlogopite mica of $90\times10^{-7}°C^{-1}$. The developed β-spodumene phase tends to harden the body and thereby impair the machinability character according to their amount, growth, and distribution in the glassy matrix. The microhardness can reach up to 1,200 MPa. When the proportion of β-spodumene solid solution approaches more than 50%, the hardness increases sharply and the ceramic is no longer considered machinable.

While the art of dental porcelain is similar to the technique of glass ceramics preparation, the porcelain term here refers to a glass ceramic made by sintering of amorphous glass frit powders under controlled crystallization, but it is of course glass ceramic made by crystallization of powder glass. The starting materials for the glass ceramics are amorphous glasses that successfully crystallize in a controlled process.

The glass ceramic can be produced by three different processes, namely:

- The process of obtaining glass by fusing a mixture of raw materials and then solidifying into glass blocks which are reheated to develop nuclei and then treated at higher temperature to develop crystal phases in the glass.
- The process of obtaining glass by fusing a mixture of raw materials, solidifying by quenching into glass frits and milling the frits into fine powders which are

then pressed, and heat treated to achieve full sintering and develop crystal phases in the glass matrix.

- The process of obtaining amorphous glass powder by subjecting a gel based on metal alkoxide compounds, to a high-temperature heat treatment. The powder is then ground, shaped, and heat-treated to form crystals in the amorphous glass matrix.

Because glass-ceramics have a crystallinity content generally greater than 50 vol.%, they are normally mechanically stronger than their parent glasses. Hence, annealed glass bodies conventionally demonstrate a modulus of rupture (MOR) value in the range of about 35–60 MPa, whereas glass-ceramic products created from these glasses will exhibit moduli of rupture in the range of 70–300 MPa. Although the latter values represent a significant improvement, numerous investigations have been undertaken to enhance the mechanical strength of glass-ceramic bodies even further.

In general, the crystallization process is time–temperature related. Therefore, only short exposure periods will be required at the higher temperatures, e.g., half an hour or even less; whereas much longer periods of time will be employed at lower temperatures, e.g., up to 24 h or more.

7.2 Mechanisms of Nucleation and Crystallization

The crystallization is the process by which a regular crystalline array is generated from the less ordered glass structure. The crystallization of a glass seems to start from distinct centers and crystal growth takes place by the deposition of materials upon the tiny particles (nuclei).

The crystallization process includes two distinct processes: nucleation and crystallization. Nucleation is a key factor in the controlled crystallization of a glass ceramic as it involves the formation of regions of longer range atomic order than the amorphous glass matrix, representing unstable intermediate states. These regions are capable of developing spontaneously into a stable crystalline phase.

Frequently, the glass body is exposed to a two-stage heat treatment. In this treatment, the glass needs to be heated initially to a temperature within or somewhat above the transition range for a period of time sufficient to induce the development of sufficient nuclei in the glass. Thereafter, the temperature needs to be increased to levels approaching or above the softening point of the glass to induce the growth of crystals upon the nuclei. The general manufacturing schedule for the production of glass ceramics starting from glass melting is shown in Fig. 7.2.

The nucleation can be classified as either homogeneous or heterogeneous. In homogeneous nucleation, the nuclei are chemically of the same materials as the growing crystals. The process of homogeneous nucleation of most glasses is difficult and tends to be impossible because of the hazards of uncontrolled crystallization, which is commonly referred to as devitrification.

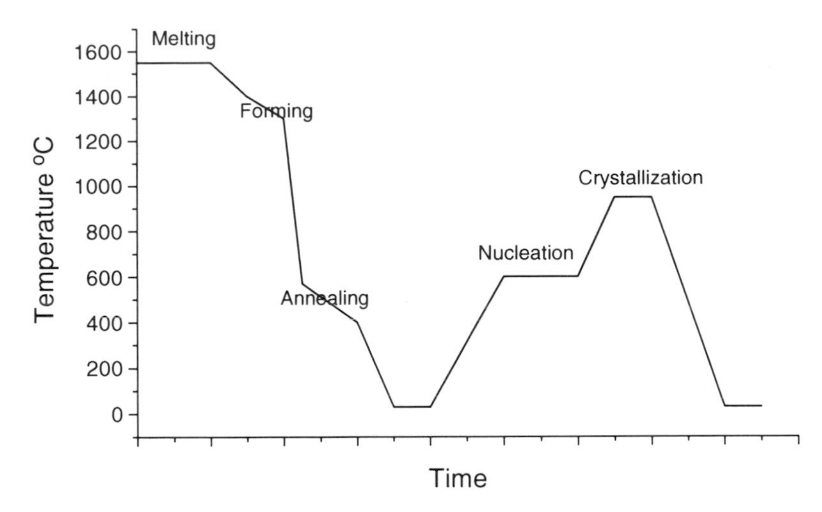

Fig. 7.2 Processing diagram for the formation of a glass ceramic

In heterogeneous nucleation, the nuclei are quite different chemically from the developed crystals and are created through the use of nucleating agents such as TiO_2 or ZrO_2. The nucleation can start from the surface of the glass or from within the bulk of the glass. Also, the nucleation can be catalyzed to control the crystallization process of the starting glass to achieve the desired glass ceramics.

Liquid–liquid phase separation is another way that is employed in many compositions during the crystallization of glasses into glass ceramics. Liquid–liquid phase separation is a common phenomenon in multicomponent oxide glasses, such as those employed in making mica glass ceramics, and occurs at temperatures below the liquidus temperature, even at relatively high viscosities. The newly formed glassy phases develop inhomogeneous structures, consisting of either interconnected phases or isolated rounded regions embedded in a continuous matrix glass (the so-called droplet structure). The type of micro- or nanostructure formed depends on the composition of the original glass, temperature, and dwell time of thermal treatment.

Particular glasses have a tendency to phase separate into silica-rich glassy matrix and an alkaline-rich glassy droplet phase. Examples of the glasses, which show this type of behavior, are those in the systems SiO_2–Al_2O_3–LiO_2 and SiO_2–Al_2O_3–MgO, making it possible to control the nucleation of the crystalline phase by careful control of the phase separation process.

The nucleation process of fluorine mica phases in the glass matrix is also known to develop by liquid–liquid phase separation. Upon cooling of the glass melt, some glasses tend to undergo spontaneous liquid–liquid phase separation as shown in Fig. 7.3, giving rise to isolated rounded regions which results in white opalization. The opalization is due to the formation of fluorine-rich droplets resulted by liquid–liquid phase separation through composition modification and heat treatment.

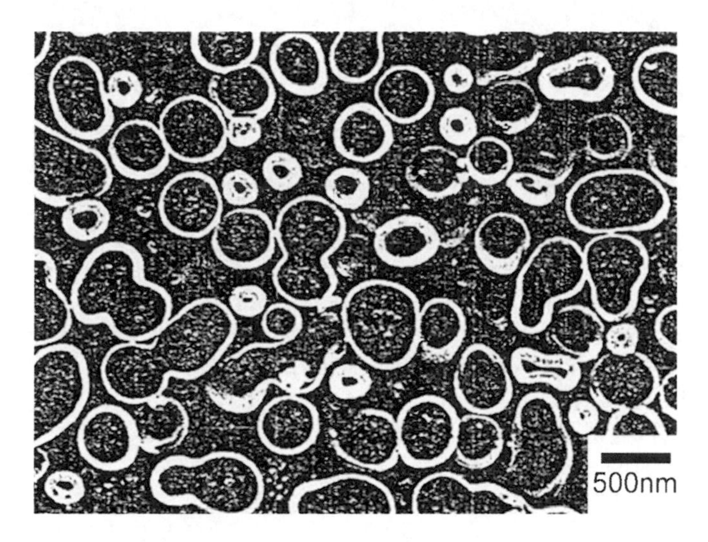

Fig. 7.3 Liquid–liquid phase separation

7.2.1 Bulk/Volume Nucleation

Bulk crystallization is the most desirable crystallization mechanism and is used in the manufacture of almost all commercial glass ceramics as it assures a uniform microstructure with high crystallinity content. This means that the product can be created in its desired final glass shape and then heat treated to form the crystal structure of the glass ceramics.

The glass ceramic process generally comprises the preparation of parent glass by melting the proper ingredients and then the glass crystallizes into microcrystalline glass ceramics through controlled heat treatment.

The ideal crystallization procedure involves two-step a process which helps to induce substantial nucleation and that insures rapid and uniform subsequent crystallization. First, the glass is heated to a temperature, somewhat above the transition temperature of the glass, and held within that transition range for a sufficient period of time to assure good nucleation. Subsequently, the nucleated body is heated to a temperature within the crystallization range for a time sufficient to complete crystal growth. The presence of a nucleating agent is the powerful driving force for nucleation in most glass compositions.

The rate of crystal growth depends on the temperature. At temperatures slightly above the transition range, crystal growth is quite slow which may expose the glass body to deformation. Therefore, the rate of heating for the glass at temperatures above the transition range should be sufficiently slow to provide adequate time to induce sufficient growth of crystals, enough to support the body against deformation. Heating rates of 10°C/min and higher can be successfully employed, where physical support for the parent glass is provided to prevent deformation. However, heating rates of about 3–5°C/min are preferred since this can yield glass ceramics exhibiting very little or no deformation. Following annealing, exposing the glass to heat treatment schedules at the rate of about 5°C/min to the recorded dwell temperature is an acceptable schedule.

7.2.2 *Surface Nucleation and Crystallization*

Whereas the majority of glass ceramics rely on volume (bulk) nucleation, there are many glasses that cannot nucleate by this mechanism. For glass bodies that are to be shaped in the glassy state and then crystallized by heat treatment, surface nucleation and crystallization is not suitable. Only in some special cases is surface crystallization used, usually to produce a low expansion compressive layer on the glass surface in order to improve the surface properties such as the surface hardness or toughness. An example of this approach is for some lithium aluminosilicate glasses, where only the surface of the material crystallizes by surface nucleation and the interior remains glassy. Ideally, the surface phase has a slightly lower thermal expansion than the glass to generate compressive stresses and a refractive index similar to that of the parent glass.

In the case of surface nucleation, a crystallized glass ceramic with satisfactory physical properties is more difficult to achieve. Surface crystallization is not as reliable as bulk crystallization except in limited compositions. However, there are a number of glasses that are used in dentistry that rely on surface nucleation such as the leucite forming glasses in the system SiO_2–Al_2O_3–K_2O.

7.3 Selection of Glass Compositions for Glass Ceramics Processing

There are several criteria that should be considered in the selection of glass compositions for glass ceramic processing. The glass must be capable of being crystallized without the use of a long schedule of heat treatments. The constituents of the glass and their proportions must be chosen to encourage the development of a certain crystalline phase and produce the desired characteristics. For example, the potassium aluminosilicate glasses, especially the stoichiometric leucite compositions are very difficult to crystallize. Thus, glasses of such compositions are of very high stability, poor crystallinity, and insufficient crystal growth and should be avoided or modified when seeking to fabricate glass ceramics.

On the other hand, in glasses that contain a fairly high proportion of modifying oxides, these oxides weaken the glass network structure by introducing non-bridging oxygens, which link to the adjacent silica tetrahedra. As the proportion of non-bridging oxygens increase, the network structure becomes increasingly destabilized. However, the atomic rearrangement, which is so necessary for the proposed crystallization, is made possible. Li_2O, Na_2O, and K_2O are useful modifying

Table 7.1 Selected initial glass compositions of mica–cordierite glass ceramics in wt% (Albert et al. 1988)

Oxides	SiO_2	Al_2O_3	MgO	Na_2O	K_2O	F^-	CaO	P_2O_5
1	45.2	29.6	12.0	3.9	4.5	4.2	0.1	0.3
2	47.3	25.7	11.1	3.1	4.8	3.5	0.5	4.0
3	43.6	26.3	14.7	5.2	4.3	4.5	0.1	1.1

constituents during the glass ceramic crystallization process. Alternatively, the alkaline earth oxides, such as ZnO, MgO, and CaO, are useful for enhancing the crystallization process.

A range of glass compositions, shown in Table 7.1, can be used to develop mica–cordierite glass ceramics by heat treatment at temperature between 650 and 1,020°C. The micaceous cordierite glass ceramic shows a fracture toughness (K_{IC}) of up to 2.0 MPa $m^{1/2}$, hardness of 300–1,000 HV, a compressive strength of up to 450 MPa, and linear thermal expansion coefficient of $75–125 \times 10^{-7}$/K. The glass ceramic demonstrates a good chemical stability and excellent machinability as well as a high resistance to wear. During the heat treatment, mica was found to form first followed by the formation of cordierite that makes up 5–30 vol%. The grain size of cordierite is of the order of 0.5–5 μm, whereas the grain size of mica is much larger at 10–200 μm (Albert et al. 1988). The mica–cordierite glass ceramic compositions provide excellent biocompatibility and can be used for biomedical applications, such as inlays, crowns, bridges, and as veneer laminates for veneering metal frameworks or natural teeth.

7.4 Optimum Heat Treatment Conditions

The crystallization of glass into glass ceramics is normally guided by the thermal analysis data. The heat treatment converts the glass objects into uniform microcrystalline glass ceramics with superior properties compared with the parent glass. In addition, the development of a finely divided uniform microstructure or interlocking microstructure results in improvements in the strength and encourages the development of a range of desirable glass ceramic properties. The factors that control crystallization are the heating rate, nucleation temperature, crystallization temperature, and the length of heating time.

Efficient nucleation leads to the production of large numbers of small crystals rather than small numbers of large crystals, which is the main target of the glass ceramic process. Then, it is necessary to elevate the temperature of the nucleated glass in order to permit crystal growth upon the developed nuclei. This must be done at a carefully controlled rate of heating in order to avoid the deformation of the glass ceramic during the crystallization process. If the rate of heating is too high, the rate of crystal growth may not be sufficiently rapid and this could result in glass deformation. In contrast, if the rate of heating is low, deformation is less likely to happen as the remaining glassy phase decreases progressively with temperature elevation.

Determination of the transition temperature is critical to the optimization of the heat treatment schedule. The transition temperature (T_g) is the temperature at which the molten glass converts to an amorphous solid. The optimum nucleation is estimated to be close to the annealing point of a glass, so the glass is supposed to be nucleated by heating the glass to a temperature slightly higher than the glass transition temperature. The rate of heating during nucleation is not as important as during annealing or crystallization.

The nucleation temperature can be set at any temperature between the annealing temperature and a temperature slightly above the temperature for optimum nucleation. The optimum nucleation temperature generally exists within the temperature range between the transition temperature (T_g) and a temperature somewhat 100°C above T_g. This transition range corresponds to the glass viscosities of 10^{12}–10^{14} poises. It is important that during the nucleation, the glass is kept below the maximum nucleation temperature to avoid any signs of crystal growth.

The crystallization process should immediately follow the nucleation process. Following the nucleation stage, the temperature of the glass should increase at a heating rate, sufficiently slow to permit crystal growth to happen. The slow heating rate ensures safe maturing without introducing undesirable deformation. Generally, the rate of crystallization speeds up as the melting temperature of the predominant crystalline phase is approached. In this case, it is better for the heating rate not to exceed 5°C/min in order to prevent deformation of the glass object.

The crystallization of some glasses can be performed slightly above the nucleation temperature and in other cases, to achieve crystallization the temperature may need to be raised by more than 300°C above the nucleation temperature. The crystallization temperature for a glass ceramic is chosen such that maximum crystallization can be achieved without deformation. The temperature at which the glass ceramic starts to deform appreciably corresponds to the melting temperature (liquidus) of the predominant crystalline phase. Increasing the temperature above this liquidus temperature may cause the phase to redissolve. The upper crystallization temperature is selected such as to be lower by some 25–50°C than the temperature at which the predominant crystalline phase will redissolve.

7.5 Prediction of the Proper Glass Heat Treatment Schedule

The prediction of the proper heat treatment schedule is executed using various techniques depending on the type and nature of materials. In the current case, the proper heat treatment for the starting glass include many steps depending on many factors. The first factor is to know the mechanism of crystallization, i.e., whether it is bulk volume crystallization or surface crystallization. Then according to the type of crystallization mechanism, the heat treatment steps can be scheduled. The glass is used in the form of a powder in the case of surface crystallization or bulk glass in the case of volume crystallization.

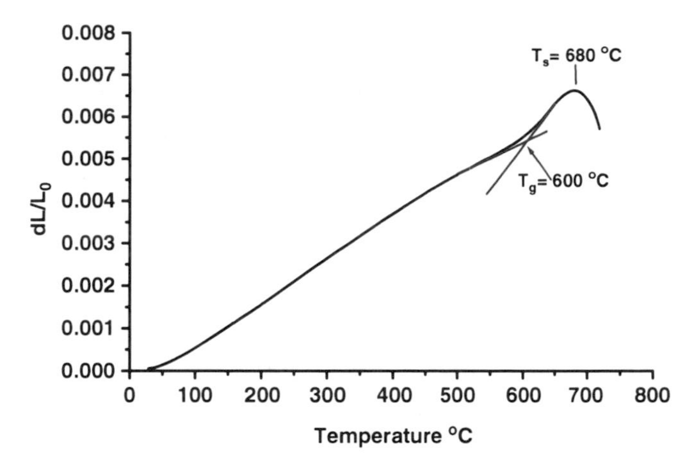

Fig. 7.4 The measurement of the transition (T_g) and softening (T_s) temperatures of a glass based on the thermal expansion curve

7.5.1 Glasses Crystallizing via Bulk Volume Crystallization

This type of glass is prepared by first melting and casting a glass block or objects from the viscous molten glass in certain shapes at high temperature followed by annealing to avoid forming stresses. The mold would need to be made for a highly refractory material such a graphite.

The heat treatment for the glass, involving the nucleation and crystallization schedule steps are guided by the DTA analysis. The effect of the developed crystalline phase(s), their relative content, the chemical composition, and the addition of mineralizers can then be assessed.

7.5.1.1 Annealing Temperature Determination

Annealing is very important as it removes as much as possible the stresses developed in the glasses during the heating or casting. It is known that nonannealed or badly annealed glasses are difficult to handle, cut or polish and in addition they can spontaneously shatter during handling due to the developed stresses, which can be very dangerous. Thus, it is important to determine the right annealing temperature of a glass before handling the glass or processing it into a glass ceramic, particularly in the case of glasses that crystallize by a volume crystallization mechanism.

The annealing temperature can be determined from either the expansion curve as shown in Fig. 7.4 or from the DTA curve as shown in Fig. 7.5. By knowing the transition temperature (T_g), it is possible to know the right annealing temperature, which lies within the annealing range between the upper and lower annealing temperatures. The transition temperature is the temperature at which a glass melt change

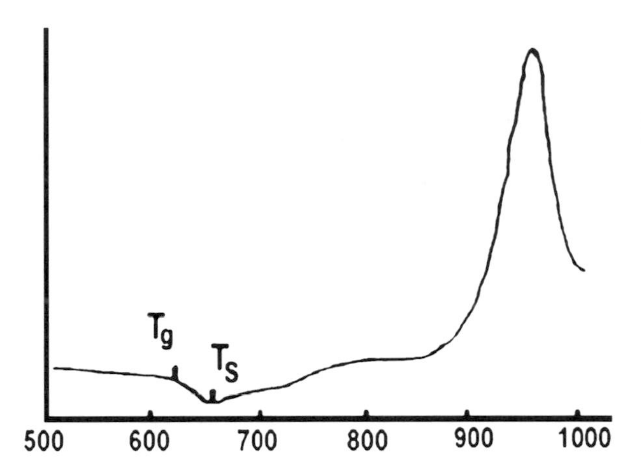

Fig. 7.5 The measurement of the transition (T_g) and softening (T_s) temperatures of a glass based on the DTA curve

into an amorphous solid and generally is estimated to be close to the annealing point of the glass. In most cases, the correct annealing step is to choose a temperature close to the transition temperature, typically within 25–50°C.

7.5.1.2 The Nucleation Temperature

The nucleation temperature of glasses is best determined from a DTA curve. The preferred nucleation is to heat the glass to a temperature above the transition temperature (T_g) for a sufficient time to assure good nucleation. In some cases, the nucleation temperature may be higher or lower in glasses with a wide transition range of temperature depending on the parent glass composition. In other words, the optimum nucleation temperature occurs between the transition temperature T_g and a temperature some 50–100°C above T_g. In addition, the nucleation temperature is determined to be close to the lowest temperature in the endothermic reaction corresponding to T_s as shown in the DTA curve in Fig. 7.5.

7.5.1.3 The Crystallization Temperature

When the nucleated glass is exposed to a secondary heat treatment at temperatures higher than the glass transition temperature (T_g), the crystallization process starts. The crystallization temperatures can be determined from the DTA curve to be at the temperature corresponding to the exothermic peak or peaks as shown in the example in Fig. 7.5. One or more phases can crystallize from the glass through exothermic reactions. The nucleated body should be heated to or slightly higher than the crystallization temperature for a time sufficient to induce crystal growth without the dissociation of the predominant crystal phase. In other words, the crystallization range lies above the nucleation temperature and a temperature slightly lower than the temperature at which the predominant crystalline phase starts to redissolve.

7.5.2 *Glass Crystallizing via Surface Crystallization Mechanism*

The glasses that crystallize via the surface crystallization mechanism from fine powders do not have a problem of annealing as they are prepared by pressing or slip molding followed by sintering at the proper crystallization temperature. Each of the glass particles in the formed glass object behaves as a separate block. The high surface area is considered to be the main driving factor for the crystallization.

Taking into consideration the previously mentioned transition, softening, nucleation and crystallization temperatures, a glass block made by pressing or formed by slip molding is sintered at a temperature in the crystallization range exceeding the softening temperature to achieve proper densification and controlled phase formation.

The combination of a high surface energy and high surface area of the glass powders drives the crystallization with no need for a nucleating agent, in contrast to a volume crystallizing glass where the nucleating agent is a vital ingredient. As the surface area of the starting glass powder increases, the position of the endothermic peak is shifted to lower temperature position in the case of glasses crystallizing by surface crystallization mechanism. This effect can be used to confirm if the system being investigated is a surface or bulk nucleating glass. By producing a set of DTA curves for powders of the same glass with reducing the particle sizes (200, 100, 45μm), if the positions of the nucleation/crystallization peaks are seen to shift to lower temperatures then the system is surface nucleating. On the other hand, if the positions of the nucleation/crystallization peaks display no significant shift to lower temperatures then the system is bulk/volume nucleating.

7.5.3 *Glasses Crystallizing by Both Mechanisms: Surface and Volume Crystallization*

Some glasses, such as miserite glass ceramics or lithium disilicate-based glass ceramics, will crystallize via either volume or surface crystallization mechanisms, depending on the chemical compositions. These glasses can potentially benefit from the combined effects of bulk nucleation by the addition of a nucleating agent and surface crystallization.

7.6 Interpretation of a Differential Thermal Analysis Curve

The chemical reactions or structural changes within crystalline or amorphous substances are accompanied by the evolution or absorption of energy in the form of heat. For example when a glass substance crystallizes, an exothermic reaction occurs since the free energy of the developed regular crystal phase is less than that of the

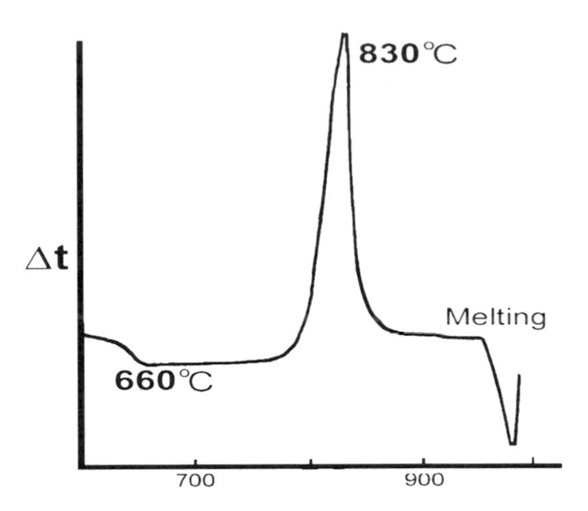

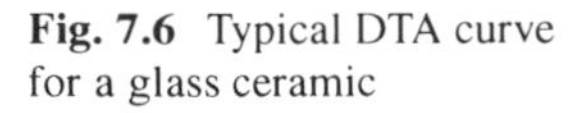
Fig. 7.6 Typical DTA curve for a glass ceramic

structurally disordered glass substance. The melting of a crystalline glass ceramics is represented by an endothermic reaction as the liquid state needs higher free energy than that of the glass.

Also chemical reactions between two materials can give rise to either endothermic or exothermic reactions. From the shape of the melting endothermic peak, it can be predicted if the glass ceramic is monocrystalline phase or multiphase. In general, the endothermic peak relating to the melting of the monophase glass ceramic is always sharp with steep dip as shown in Fig. 7.6. Thus, differential thermal analysis (DTA) is a technique that facilitates the study of phase development, decomposition, or phase transformations during the glass ceramic process with heating.

In this method the glassy materials under test is in the form of finely divided powders placed in a small crucible of platinum or other suitable refractory crucible such as aluminum oxide. Adjacent to the test crucible is a second crucible containing an inert powder such as aluminum oxide, which does not exhibit endothermic or exothermic reactions. The temperature differences between the tested sample and the reference sample powder (inert alumina) are recorded during heating and the differential temperature is plotted. The exothermic reactions are represented as peaks and the endothermic reactions are represented by dips on the DTA curve.

The technique is considered a useful method for the investigation of the glasses and the determination of the optimum temperatures at which the different crystal phases are formed. The quality and the shape of the peaks are affected by the rate of heating and the fineness of the powder.

As the temperature is increased, a dip is observed on the DTA curve due to slight absorption of heat that occurs when the annealing point of the glass is reached. With further increase of the temperature, one or more quite sharp exothermic peaks can be expected, corresponding to the appearance of various crystalline phases. At higher temperature, an endothermic reaction peak relates to the early incongruent melting of the predominant crystalline phase. Figure 7.7 shows the main features that can be determined from the typical DTA curve of a glass.

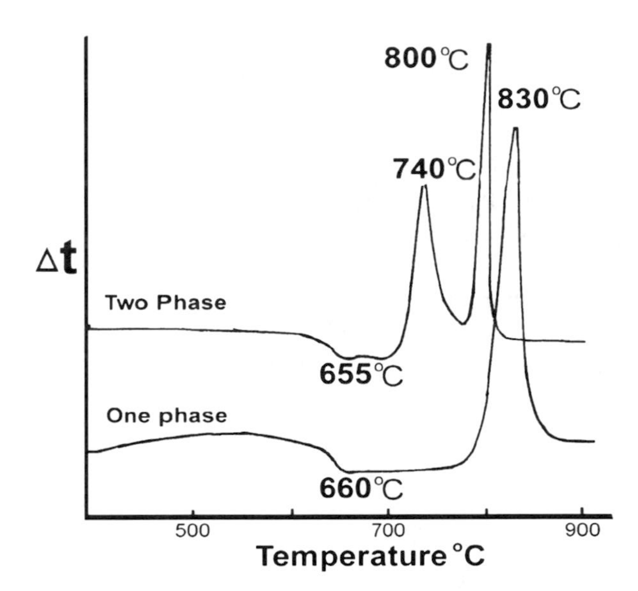

Fig. 7.7 DTA of fluorrichterite–enstatite glass ceramics

Thus, the DTA analysis yields a great deal of very useful information, which is of great help in designing the heat treatment schedules for the crystallization of a glass into a glass ceramic. The curve not only indicates the temperature range in which the nucleation and crystallization occurs, but it also indicates the maximum temperature to which the glass ceramics can be heated without exposing the glass ceramic to deformation due to melting of the crystal phases.

Having determined the exothermic and endothermic peaks, the crystallization of the different crystalline phases can be perfectly assessed as shown in Fig. 7.7. The figure shows the DTA curves of two glasses: one of them can crystallize to yield a two phase glass ceramic and the other yields a one-phase glass ceramic. In cooperation with XRD data, it is possible to assess the best schedule for the crystallization of a certain phase or phases.

One phase: fluorrichterite
Two phase: fluorrichterite–enstatite

7.7 Interpretation of Thermal Expansion Curves

The expansion curve of glass is also very important source of information in the construction of the heating schedule for the crystallization of a glass ceramic. Several significant points in the expansion curve of glass can be predicted from the thermal expansion curve of glass, including the lower annealing point, transition temperature, the upper annealing point, and the softening point in addition to the nature of the melting. The proposed points, including T_g and T_s, are very important in determining the best route to crystallization. Also, the annealing temperature of glass during preparation of glass blocks for making dental glass ceramics by CAD/CAM machine can be predicted. Figure 7.8 shows the transition temperature, softening temperature, and the thermal expansion coefficient on the thermal expansion curve of a leucite glass ceramic.

In addition, the thermal expansion of the prepared glass ceramic is very important for different applications. A low thermal expansion glass ceramic is required if one wants to improve the thermal shock resistance. A high thermal expansion coefficient, probably between 13 and 15×10^{-6}/°C, is required if the glass ceramic is used to conceal a Cr–Ni alloy with a glass ceramics layer. As can be seen, adjusting the thermal expansion coefficient helps to fix various industrial problems. Tables 7.2 and 7.3 show the thermal expansion of some glass ceramic phases.

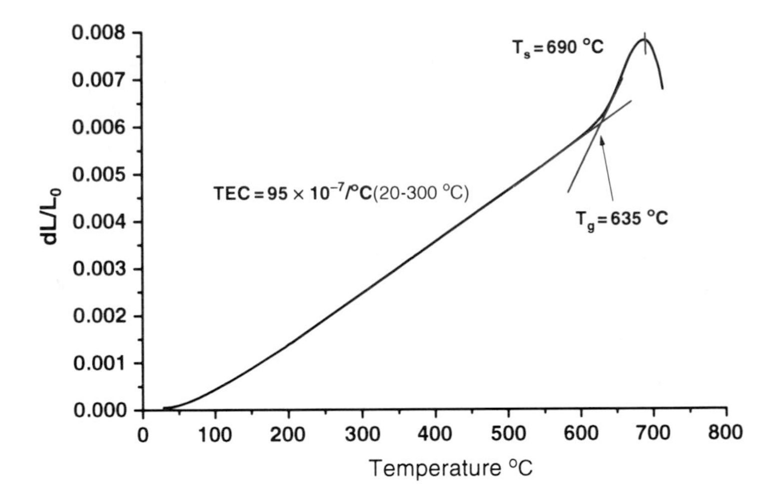

Fig. 7.8 The determination of T_g, and TEC from the thermal expansion curve for a leucite glass ceramic

Table 7.2 Linear thermal expansion coefficients of some glass ceramic phases

Crystal phase	Linear TEC ($\times 10^{-7}$/°C)	Temperature range (°C)
Β-eucryptite (Li_2O–Al_2O_3–$2SiO_2$)	−86	20–700
	−64	20–1,000
Aluminum titanate (Al_2O_3–TiO_2)	−19	20–1,000
Cordierite ($2MgO$–$2Al_2O_3$–$5SiO_2$)	26	25–700
β-Spodumene (Li_2O–Al_2O_3–$4SiO_2$)	9	20–1,000
Celsian (BaO–Al_2O_3–$2SiO_2$)	27	20–100
Anorthite (CaO–Al_2O_3–$2SiO_2$)	45	100–200
Clinoenstatite (MgO–SiO_2)	78	100–200
Magnesium titanate (MgO–TiO_2)	79	25–1,000
Forsterite ($2MgO$–SiO_2)	94	100–200

(continued)

Table 7.2 (continued)

Crystal phase	Linear TEC (×10⁻⁷/°C)	Temperature range (°C)
Wollastonite ($CaO–SiO_2$)	94	100–200
Lithium disilicate ($Li_2O–2SiO_2$)	110	20–600
Quartz silica (SiO_2)	112	20–100
	132	20–300
	237	20–600
Cristobalite (SiO_2)	125	20–100
	500	20–300
	271	20–600
	175	20–100
	250	20–200
	144	20–600

Table 7.3 The linear thermal expansion coefficients of different glass ceramics with various phase compositions

Chemical composition (wt%)						TEC	
SiO_2	Al_2O_3	MgO	Li_2O	TiO_2	P_2O_5	($\times 10^{-7}$/°C)	Crystal phases
54.5	34.5	00	5.5	5.5	00	1.1	β-Eucryptite
70.7	18.1	00	2.6	4.8	00	5.1	β-Spodumene
45.5	30.5	12.5	00	11.5	00	14.1	Cordierite, rutile
65.5	21	00	9.0	4.5	00	14.5	β-Spodumene, rutile
45.5	25.3	17.8	00	11.1	00	22.6	Cordierite, magnesium titanate
52.5	26.5	11.9	00	9.1	00	28.3	Cordierite, rutile
57.8	8.9	22.2	00	11.1	00	39.9	Cordierite, rutile, cristobalite
82.3	0.0	3.7	11	00	3	145	Lithium disilicate, quartz

Stookey (1960), McMillan and Partridge (1963), and McMillan (1964)

Severe internal stresses can be generated due to strains resulting from uneven thermal expansion of the material. For a plate shaped glass ceramic heated on one face and cooled on the other face, compressive stresses are set up on the hotter face and tensile stresses are generated on the cooler face. A temperature gradient is established across the thickness of the glass ceramics where the uniformity of the temperature gradient depends on the thickness of the sample.

A good example for the thermal expansion modification of diopside–mica glass ceramics based on the modification of the glass chemical composition is shown in Fig. 7.9 and Table 7.4.

With sudden cooling, the stresses generated are higher than those of slower cooling conditions because initially the temperature gradient is confined to a thin surface layer and the surface stress may attain a value twice that of the steady-state conditions.

The potential for failure of a glass ceramic is much higher if tensile stresses are generated as a result of differential cooling or heating and if the stresses generated exceed, the breaking stress of the material fracture will occur. The extent to which the material can withstand a sudden temperature change without fracture is referred

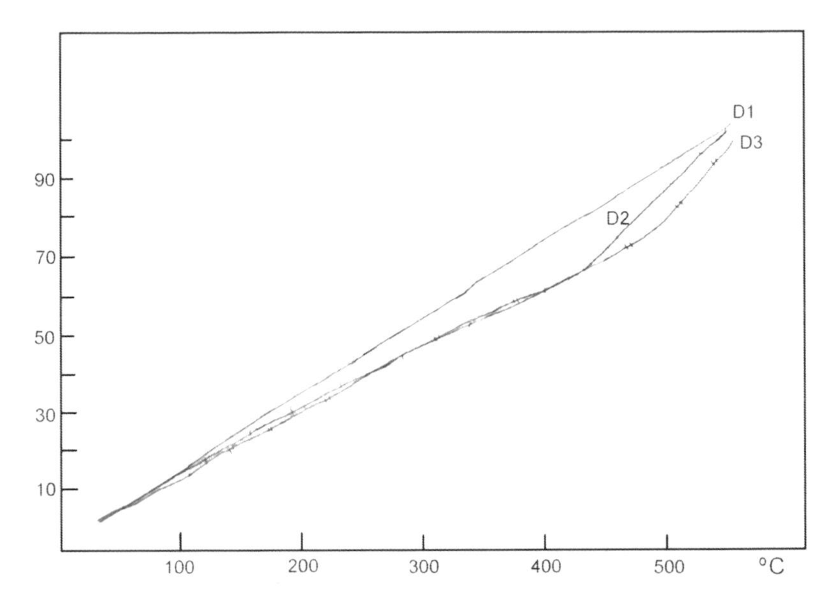

Fig. 7.9 Thermal expansion modification of diopside–mica glass ceramics

Table 7.4 Thermal expansion modification of diopside–mica glass ceramics

Sample No.	Heat treatment parameter (°C/h)	Thermal expansion coefficient (×10^{-7}/°C) 20–200°C	20–400°C	Present phases
D1	–	82	87	Amorphous
	1,000/2	100	101	Di–Ph
D2	–	99	99	Amorphous
	1,000/2	103	95	Di–Ph
D3	–	87	91	Amorphous
	1,000/2	95	91	Ph–Di

to as the thermal shock resistance. The thermal shock resistance is defined in terms of the maximum temperature interval through which the material can be rapidly heated and cooled without fracturing. The resistance to sudden cooling is thought to be more important than the resistance to sudden heating as most serious cracking occurs on cooling.

7.8 Significant Points on the Thermal Expansion Curve

Several significant points in the expansion curve can be predicted from the thermal expansion curve of a glass, including the lower annealing point, transition temperature, the upper annealing point, and the softening point in addition to the nature of the melting (Fig. 7.10).

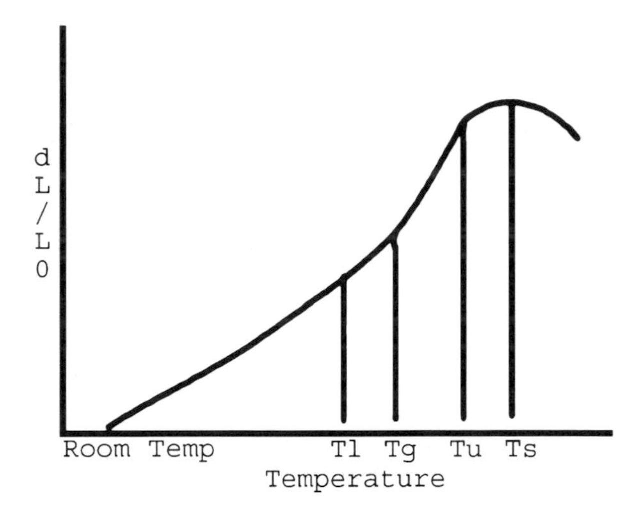

Fig. 7.10 Significant points along the thermal expansion curve of a glass, where T_l is the lower annealing or strain release temperature, T_g is the transition temperature, T_u is the upper annealing temperature, and T_s is the softening temperature

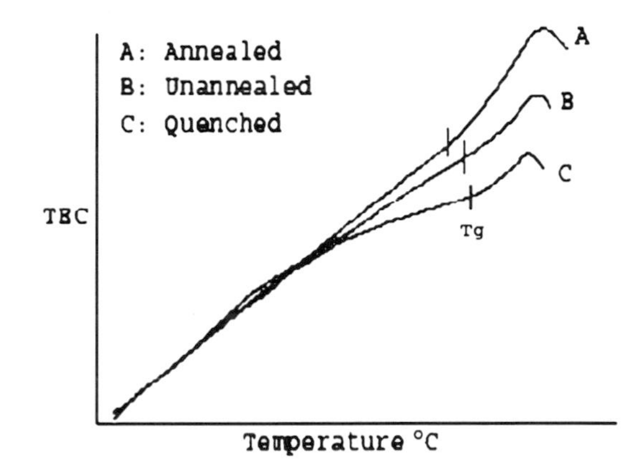

Fig. 7.11 Variation in expansion curve as a consequence of the annealing of the glass

7.9 Dependency of the TEC on Heat Treatment

The temperature dependence of the linear thermal expansion is expressed by the dilatometric curve. The expansion is shown to increase uniformly and continuously without any bends or deflections if structural changes do not occur and the glass has a good stability and is well annealed as shown in Fig. 7.11.

The dilatometric curve of a rapidly quenched or poorly annealed glass differs from that of a well-annealed glass. At lower temperatures the expansion coefficient of a quenched glass is somewhat higher than that of an annealed glass. As consequence of the difference, the expansion curve for a quenched glass shows a distinct deflection over a temperature range of 100°C leading up to T_g.

The deflection is due to the considerable shrinkage of a quenched glass that has not been annealed and increases as the transition temperature is approached. The reason for this is that the quenched glass has a relatively lower density than the annealed glass and the densification of the quenched glass is achieved prior to the transition temperature. Above T_g the quenched glass expands in the same way as the annealed glass as better densification is achieved. Thus, the transition temperatures must be determined for the glass after it has been thoroughly annealed.

7.10 Physical Changes Due to Crystallization

An obvious change brought out by the heat treatment is the conversion of the transparent glass into a translucent or opaque polycrystalline glass ceramic. The opacity of the crystalline ceramic is due to scattering of the light at the interfaces between adjacent crystals and between the crystals and the residual glass phase due to differences in the refractive index.

In some cases, where the crystals are very fine, near to the wavelength of light (390–750 nm) and the refractive index of the crystalline phase is close to that of the glass, the glass ceramic can appear to be transparent or slightly translucent. There are several crystalline phases that can give such an effect, e.g., leucite, diopside, or lithium disilicate crystals dispersed in a glassy matrix. An XRD analysis would show the appearance of a complete pattern of the leucite or kalsilite phases, with more than 10% volume crystallinity, and yet the glass ceramics is still transparent.

The properties of glass ceramics depend on the kind and the percentage of developed crystalline phases together with the composition of the glassy matrix. The crystallization or phase transformation results in a volume change, which leads to a change in the specific gravity. The specific gravity of a glass ceramic is often different from that of the parent glass because of the small volume changes that may occur during the heat treatment process. These changes may involve either a slight contraction or a slight expansion, but this would not usually exceed a 3% volume change. The volume changes in glass ceramics are as a result of the overall differences in specific gravity of the crystalline phases compared with the parent glasses. The crystalline phases formed may have higher densities than the original glass. Table 7.5 shows the densities of some glass ceramic crystal phases compared with the densities of the corresponding glasses.

On the other hand, there are some compositions that may not exhibit any volume change due to the conversion of the glass to a glass ceramic. The net volume change

Table 7.5 The densities of some glass ceramic crystal phases compared with the densities of the corresponding glasses

Crystal phase	Formula	Density of glass ceramic (g/cm^3)	Density of glass (g/cm^3)
Wollastonite	$CaO{\cdot}SiO_2$	2.92	2.87
Anorthite	$CaO{\cdot}Al_2O_3{\cdot}2SiO_2$	2.75	2.60
Li-Disilicate	$Li_2O{\cdot}2SiO_2$	2.45	2.42
Cordierite	$2MgO{\cdot}2Al_2O_3{\cdot}5SiO_2$	2.53	2.47
β-Eucryptite	$Li_2O{\cdot}Al_2O_3{\cdot}2SiO_2$	2.67	2.43
Diopside	$CaO{\cdot}MgO{\cdot}2SiO_2$	3.27	2.75

Table 7.6 Chemical composition of glass batches of nepheline spodumene glass ceramics

Chemical composition (wt%)					Expansion behavior					
					Glass			Glass ceramics		
SiO_2	Al_2O_3	CaO	Na_2O	LiF	T_g (°C)	T_s (°C)	TEC	°C/2 h	TEC	Phases
41.42	35.14	9.66	10.87	2.91	562	617	84	900	29	Nepheline-β-spodumene
40.25	34.15	9.39	10.57	5.65	522	563	92	900	17	Nepheline-β-spodumene

Table 7.7 The thermal behavior of nepheline-containing glass ceramics

Chemical composition (wt%)					Expansion behavior					
					Glass			Glass ceramics		
SiO_2	Al_2O_3	CaO	Na_2O	TiO_2	T_g (°C)	T_s (°C)	TEC (×10^{-7}/°C)	°C/2 h	TEC	Phases
42.66	36.19	9.95	11.20	0.00	706	740	73	900	70	Nepheline
41.5	35.2	9.7	10.9	2.9	673	716	67	900	84	Nepheline
40.3	34.2	9.4	10.4	5.7	698	731	82	900	85	Nepheline

TEC coefficient of thermal expansion

in these compositions may be zero when the volume change made by one crystal phase is balanced by the effect of volume change by another phase. For example, the crystallization of a leucite tetragonal phase in the glass will result in a high expansion of the material, where the tetragonal leucite exhibits a TEC in the range of 20–30 × 10^{-6}/°C. In contrast, the crystallization of cubic leucite phase will result in a very low volume change. Also, the crystallization of β-spodumene or β-eucryptite in a glass ceramics will result in a very low expansion.

A good example is the simultaneous crystallization of both spodumene and nepheline in a glass to yield a glass ceramics based on the glass composition shown in Table 7.6. Spodumene–nepheline glass ceramics are prepared from Na_2O–CaO–Al_2O_3–SiO_2 glass compositions. The addition of LiF changes the course of the reaction toward the formation of spodumene together with nepheline. The crystallization of spodumene strongly reduced the thermal expansion coefficient. The addition of 3 wt% LiF is considered optimal as it reduces the expansion coefficient and enhances the uniformity of the microstructure. A higher content of LiF impairs the mechanical machinability of the glass ceramics, particularly when the content of β-spodumene tends to exceed 50%. Li^+ substitute for Na^+ cations, resulting in the formation of β-spodumene.

The presence of fluorine cations is found to promote the growth of nepheline. The transformation temperature (T_g) and the softening temperature (T_s) are reduced as a result of the addition of LiF. Another frequently seen change in the glass after crystallization is the refractoriness that increases as a result of the crystallization of nepheline in the glass.

On the other hand, the crystallization of glasses based on the SiO_2–Al_2O_3–CaO–Na_2O system with or without TiO_2 results in the crystallization of the nepheline phase in the glass ceramic as shown in Table 7.7.

Table 7.8 Intraoral conditions

Mastication load ranges	6–130 N	Temperature range	5–65°C
Maximum load ranges	200–800 N	pH range	0.5–8.0
Chewing cycles/day	1,000–1,400		

Nevertheless, the most striking and useful change achieved by the crystallization of the glass into a glass ceramic is the increase in the mechanical strength. The MOR of most glasses is of the order of 60 MPa, while the MOR of a glass ceramics can be in the range of 70–300 MPa and may even reach higher values.

7.11 The Impact of Environment on Choosing the Right Glass Ceramic

In order to be able to design the microstructure and optimize the glass ceramic for use in medical ceramics, for example, dental ceramics, it is important to understand the conditions in the mouth. Certain aspects and conditions of the intraoral environment are listed in Table 7.8. These conditions need to be considered when defining the characteristics of the ceramics that can be implanted in the oral cavity.

During mastication, the designed dental ceramic structure must withstand average cyclic loads of approximately 150–300 MPa in a moist environment. Also, the ceramic should have excellent thermal shock resistance up to 100°C and a low chemical solubility.

The ceramics to be used in the oral cavity must perform reliably under these aggressive conditions. In the case of restorations made solely out of a ceramic and to be used as posterior crowns, for example, these would need to have a mechanical strength of around 200–300 MPa, a fracture toughness of about 3–4 MPa $m^{1/2}$, and a chemical solubility of less than 100 μg/cm^2 when directly exposed to the oral environment. Thus, one possible solution is to search for mineral phases to be crystallized in glass ceramic to lend these glass ceramics the appropriate characteristics. Various mineral phases such as leucite, lithium disilicate, and fluormica constituting glass ceramics are currently used commercially in the market with varying degrees of success.

7.12 Chemical Solubility of Glass-Ceramics

Unlike the other properties, the chemical solubility is a very complex phenomenon. The chemical solubility of a glass ceramic is a function of the following factors:

- Its crystalline structure
- The chemical composition of the glass ceramic

- The surrounding environment (temperature, pH, and time of contact)
- Specimen surface area
- The surface roughness
- Production factors such as contact with a mold during shaping and annealing atmosphere
- Phase separation
- Development of low solubility crystalline phases such as canasite

The chemical components of the glass can affect the chemical solubility of glass ceramics in three ways

- The lower the heat of hydration of the oxide, the better the chemical solubility of the glass. ZnO is the oxide with the least heat of hydration, so it substantially improves the chemical solubility, particularly with respect to water vapor.
- The higher the volatility of the oxide such as K_2O, the better it is able to reduce the chemical solubility.
- The higher the degree of M–O ionic bonding, the lower the chemical solubility, the lower the covalence degree.
- The effects of water and mineral acids on the chemical solubility are actually of the same order of magnitude. Glass is attacked significantly by alcohols, strongly by steams and very corrosive with HF and H_3PO_4 and alkaline solutions such as hydroxides.
- The chemical solubility is diminished by the presence of inhibitors of alkaline corrosion such as Al, Zn, Be, Ca, Sr, Zr, and Mo.

Silicate glasses are among the most chemically inert commercial materials. Because of the chemical inertness of glasses, the chemical solubility of glasses is concerned almost entirely with their reactivity with water, aqueous solutions, and water vapor, which is extremely slow.

The reaction of alkali silicate glasses with water is complicated and involves at least two different steps: ion exchange of alkali ions in the glass with hydronium ions (H_3O^+) from water and dissolution of the glass into the liquid water. The result is that the rate of attack on silicate glass by water depends on pH, volume of solution in contact with the glass, solution concentrations, and glass composition. The first step in the reaction of water with an alkali silicate glass is the exchange of alkali ions in the glass with hydronium ions from the water.

$$Na^{+}(g) + 2H_2O = H_2O^{+}(g) + Na^{+} + OH^{-}$$

In contact with water, sodium hydroxide dissolves in the water and increases its alkalinity, if the volume of solution is small compared to the surface area of glass the pH increases rapidly. Whereas on contact with water vapor, the sodium hydroxide stays on the glass surface. It rapidly reacts with carbon dioxide from the atmosphere forming sodium bicarbonate crystals on the glass surface. In aqueous media, the surface of the glass dissolves in the solution by breaking the silicon–oxygen network and finally forming silicic acid.

The silicic acid, H_4SiO_4, is somewhat soluble in water and if the volume of solution is small compared to the surface area of glass, the solution becomes rapidly saturated with silicic acid, which can reduce the rate of dissolution.

Generally, the chemical durability is affected by several factors. As the amount of alkali in an alkali silicate glass is increased, the rate of reaction with water increases. Nevertheless, there are exceptions and the addition of alkaline earth oxides such as CaO or MgO improves the chemical durability. The addition of 5–10% CaO to sodium silicate glasses increases the chemical durability. Equally, the presence of potassium oxide in sodium silicate glass improves the durability. On the other hand, phase separation in a glass reduces the chemical durability. So it is very important to avoid phase separation in glasses for structural and chemical purposes.

7.13 The Chemical Solubility of Canasite

Although glass ceramics containing canasite as the sole phase have a high strength and toughness and good microhardness, a severe problem is the low chemical durability. For a glass ceramic to be used in the construction of a ceramic core, it must have a chemical solubility of less than 2,000 μg/cm^3. If the glass ceramic is to be used as a veneering ceramic it must have a chemical durability of less than 100 μg/cm^3. Unfortunately, the chemical solubility of canasite glass ceramics is generally found to exceed 2,000 μg/cm^3. Although some recent research has been able to bring this figure down to around 700 μg/cm^3 it is still not enough for veneering ceramic.

The chemical solubility is found to be affected by the chemical and the mineral phase composition. The predominant crystal phase developed in situ appears to be canasite, $Ca_5Na_4K_2[Si_{12}O_{30}]F_4$ with a probable solid solution to $Ca_5Na_3K_3[Si_{12}O_{30}]F_4$. The difference between the two phases is the difference in the Na_2O/K_2O ratio and probably the reason of low chemical durability is the increase in the Na_2O/K_2O ratio, which results in an increase of the glassy phase content.

One way for resolving the problem is the simultaneous crystallization of an appropriate phase with good chemical resistance and good biocompatibility such as apatite or leucite. Thus a glass ceramic, in which the fluorcanasite primary crystal phase crystallizes together with a fluorapatite secondary crystal phase in the residual glassy phase or matrix, results in improving the chemical durability. F-apatite has the formula $Ca_5(PO_4)3F$ as fluorine substitutes for the hydroxyl group conventionally present in naturally occurring apatite. The glasses can be internally nucleated depending on the presence of CaF_2 and P_2O_5. Crystallization may be achieved at time–temperature cycles varying from 550°C/4 h to 950°C/1 h.

Another practical route is to increase the content of SiO_2 up to 65% and increase the content of K_2O and CaF_2. The change is found to improve the chemical durability, but the starting glass has a tendency to devitrify due to the separation of white amorphous silica in the glass. The devitrified silica is successfully dissolved through the addition of 1% ZnO. So the canasite ceramics can have very high mechanical strength, high fracture toughness, and very good chemical resistance.

Further Reading

Albert, C., Beleites, E., Carl, G., Grosse, S., Gudziol, H., Hoeland, W., Hopp, M., Jacobi, R., Jungto, H., Knak, G., Kreisel, L., Musil, R. Naumann, K., Vogel, F., Vogel W.: Micaceous-cordierite-glass ceramic. US patent 4789649, 1988

Appen, A.A.: The Chemistry of Glasses. Himiq, Leningrad (1974) (in Russian)

Beall, G.H.: In: Hench, L.L., Freiman S.W. (eds.) Advances in Nucleation and Crystallization of Glasses, pp. 251–261. American Ceramic Society, Columbus, OH (1971)

El-Meliegy, E.M.: Machinable spodumene – fluorophlogopite glass ceramics. Ceram. Int **30**(6), 1059–1065 (2004)

El-Meliegy, E., Hamzawy, E.M.: Celsian – fluorophlogopite porcelain based on Egyptian talc. Adv Appl Ceram **140**(2), 92–96 (2005)

Hamzawy, E.M.A., El-Meliegy, E.: Crystallization in the Na_2O-CaO-Al_2O_3-SiO_2-(LiF) glass compositions. Ceram Int **33**(2), 227–231 (2007)

Hamzawy, E.M.A., El-Meliegy, E.: Preparation of nepheline glass-ceramics for dental applications. Mater. Chem. Phys **112**(2), 432–435 (2008)

McMillan, P.W.: Glass Ceramics. Academic, London and New York (1964)

McMillan and Partridge, Lithium zinc silicate glass-ceramics containing high proportions of zinc oxide were first reported by in 1963

Mustafa, E.: Fluorophlogopite porcelain based on talc feldspar mixture. Ceram. Int. **27**(1), 9–14 (2001)

Omar, A.A., Hamzawy, E.M.A., Farag, M.M.: Crystallization of alkali-modified fluorcanasite glasses. Silicate Ind **73**(5–6), 89–94 (2008)

Stookkey, Method of making ceramics and product and thereof. US patent 2920971, 1960

Stookey S.D., Controlled nucleation and crystallization lead to versatile new glass ceramics, Cand E News **39**(25), 116-25 (1961)

Volf, M.B.: Mathematical Approaches in Glasses. Elsevier, Praha (in Czech) (1984)

Chapter 8
Microstructural Optimization of Glass Ceramics

8.1 Ceramic Microstructures

The microstructure of ceramics is the most important property affecting the other ceramic properties and the application target. The microstructure is likely to be responsible for the most valuable properties of ceramics. The microstructure of a glass ceramic consists of a fine-grained phase or phases dispersed in a glassy matrix (glassy phase). Sometimes there may also be some very small pores in between grain boundaries as shown in Fig. 8.1.

Several structural and morphological elements should be taken into consideration when discussing the microstructure of glass ceramics. The grains making up the glass ceramic microstructure are the primary crystalline phase or phases. The grain boundaries represent the interface that separates two grains at which the orientation of the crystal lattice changes from one grain to another. This contrasts with a interphase boundary, which represents the boundary between grains of different chemical composition.

The morphology of the primary phase grains is characterized by describing the shape, size, and distribution of the grains dispersed in the glassy matrix, which in turn control the ceramic properties. Examples of properties that can be tailored by the optimization of the microstructure are the mechanical properties, chemical resistance, microhardness, and optical properties. The most desirable microstructure is generally described as having a fine-grained primary phase or phases with nearly the same grain size, uniform or narrow grain size distribution and random distribution of the different grains in the glassy matrix.

E. El-Meliegy and R. van Noort, *Glasses and Glass Ceramics for Medical Applications*, DOI 10.1007/978-1-4614-1228-1_8,

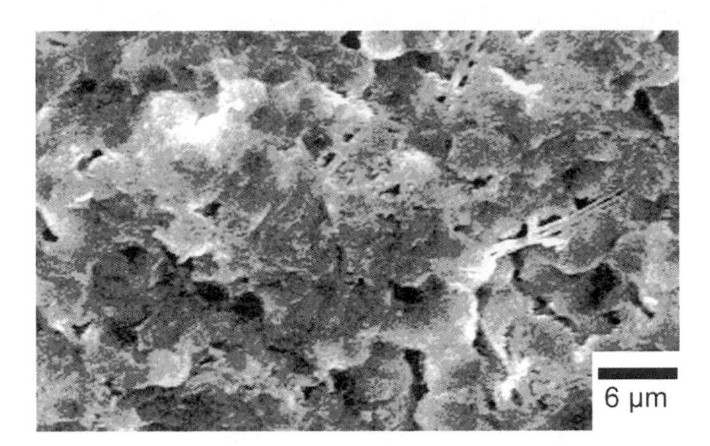

Fig. 8.1 Pores present in the microstructure between grain boundaries in a diopside glass ceramic

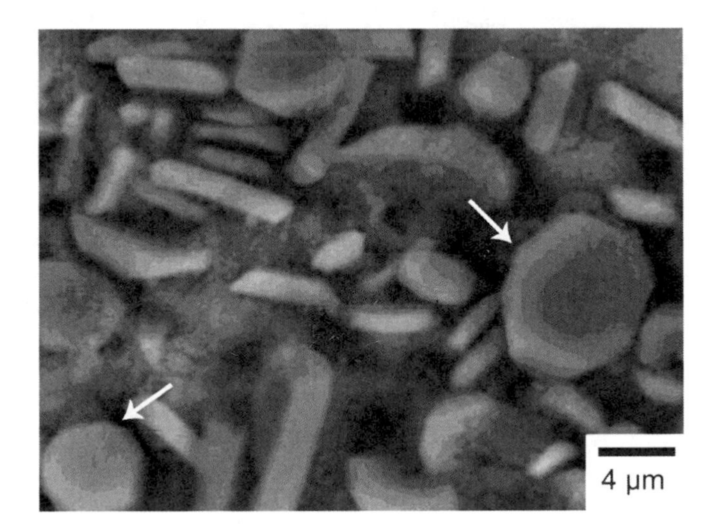

Fig. 8.2 Tabular grains of nepheline solid solution crystallized from glass containing 6% TiO_2 and heat treated at 900°C/2 h

8.1.1 Crystalline Shapes, Forms, and Habits

The grains dispersed in the matrix can have different shapes, forms, and patterns in relation to other grains, grain boundaries, and the matrix and their grain size is determined by their dimensions. The grains can be tabular as shown in Fig. 8.2, rounded (equiaxed) as shown in Fig. 8.3, platelets as shown in Fig. 8.4, columnar

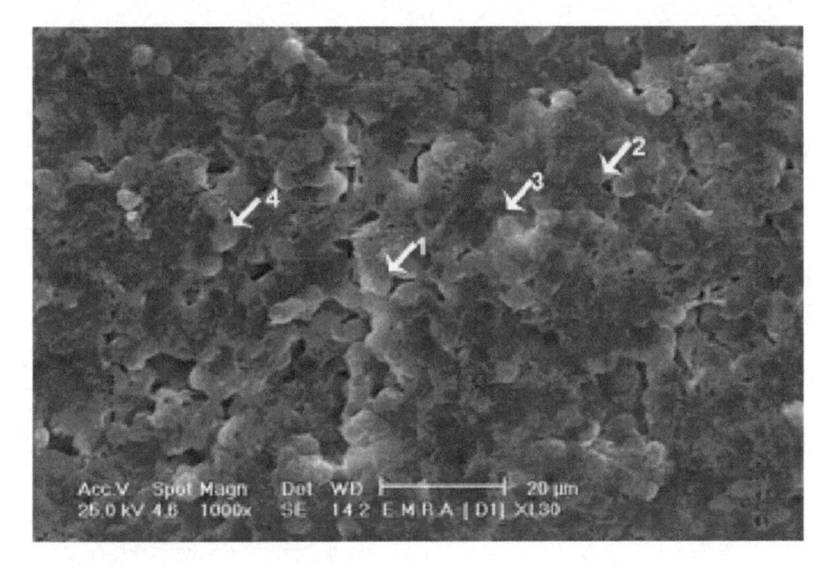

Fig. 8.3 SEM showing pores in grain boundaries (*arrows* 1 and 2) and equiaxed rounded diopside grains (*arrows* 3 and 4)

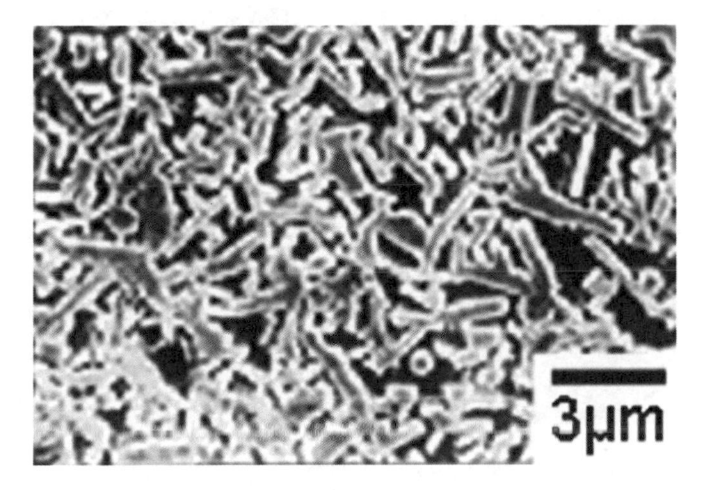

Fig. 8.4 Platelet mica grains

and euhedral as shown in Fig. 8.5, acicular (needle like), lath like as shown in Fig. 8.6, acicular shape as shown in Fig. 8.7.

Grains with asymmetrical dimensions are described according to their aspect ratio, which is defined as the ratio of the longest grain dimension to the shortest dimension. Platelets are grains with straight boundaries and a size of less than 10 μm. Tabular indicates tablet-like grains and their grain size is described by the

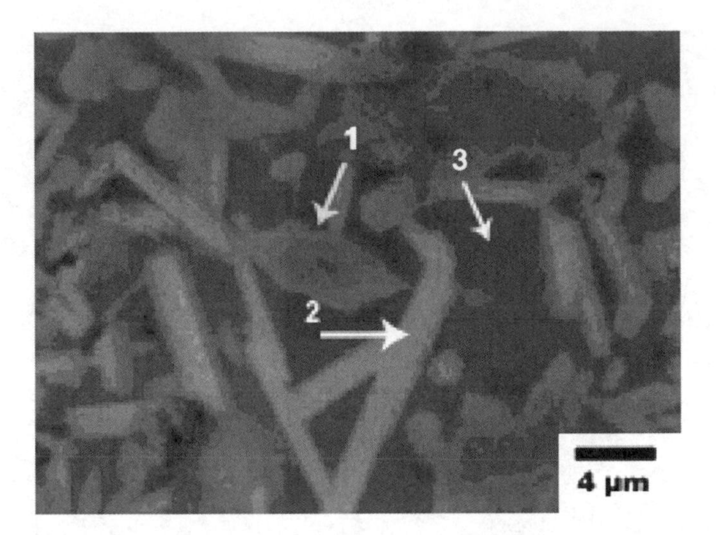

Fig. 8.5 SEM showing cross section of euhedral crystal (*arrow* 1), fluorophlogopite crystals, columnar fluorophlogopite crystals (*arrow* 2) and fine grained spodumene matrix (*arrow* 3)

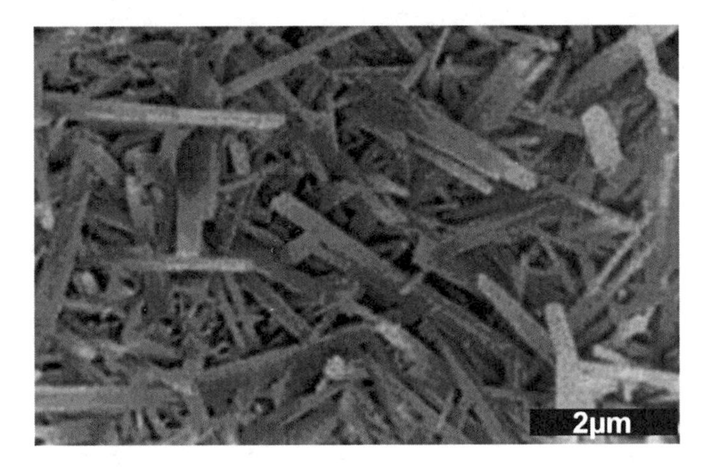

Fig. 8.6 SEM of lath like canasite glass ceramic crystals

aspect ratio. Columnar grains are formed by unidirectional grain growth. The grains can also be described as euhedral, subhedral, or anhedral to indicate the flatness of their faces.

The microstructure of a glass ceramic is best described by the pattern made by several parameters including grain morphology, the grain size and its distribution, the grain boundaries, and the associated pore morphology. The relation between the different structural and morphological element describes the ceramic microstructure.

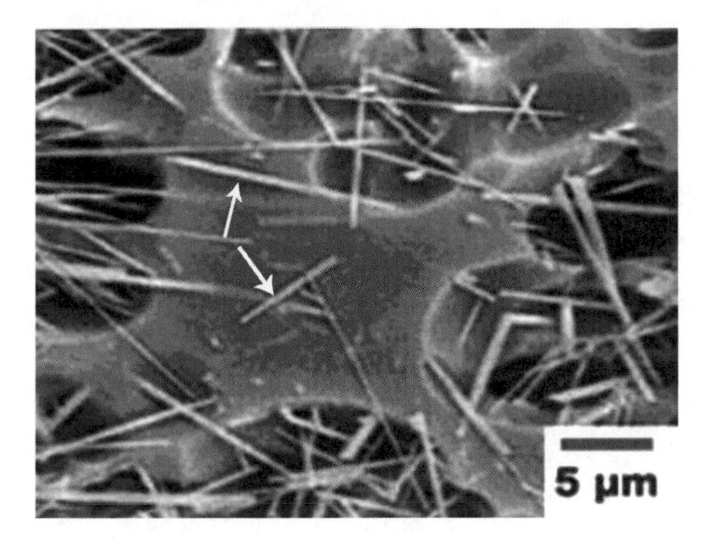

Fig. 8.7 Acicular grains in leucite glass ceramics

8.2 Development of Glass Ceramic Microstructures

In the dental application of glass ceramics, the starting point may be a glass or glass ceramic in the form of a block for CAD–CAM processing, a pellet for hot pressing or as a powder that is sintered. The blocks for CAD–CAM processing may be fully crystallized or, partially crystallized and once the shape has been machined a final crystallization firing cycle is executed. The pellets used for hot pressing are generally fully crystallized and it is important that the material has good flow characteristics to be able to fill the mold. In the case of ceramic powders, commonly referred to as a frit, the material is either a glass, e.g., feldspar, or a glass ceramic, e.g., leucite-reinforced feldspar, depending on its application.

The mechanism of densification in the case of glass powder is sintering via a viscous liquid phase (viscous phase sintering). Generally speaking, the powder will already be fully crystallized, although it is possible for the crystallization of the mineral phases to occur at the same time as the densification process.

As the route to producing a crystalline glass ceramic is adopted, it is important that the steps of nucleation, crystallization, and grain growth are well controlled such that a microcrystalline ceramic can be achieved. The control of the microstructure is dependent on the chemical composition, mineralizers, and heat treatment schedules. If the process is not carried such that the above-mentioned parameters are not controlled, then it is likely that local or abnormal grain growth will appear. With such abnormal crystal growth this can result in cracking and strength degradation. One of the most important issues is to choose the right mineralizers for developing the appropriate phases together with adjusting the rate of heating and cooling. Also, the nature of the starting glass powders as the fineness and uniformity strongly affects the microstructure, and the role of open and closed pores is significant.

8.3 Adjustment of Microstructure

The real merit of a glass ceramic is that under carefully controlled processing conditions it can have a fairly fine grain size microstructure and can achieve excellent uniformity of grains in the glassy matrix in addition to suitable grain size distribution. Once a glass has been manufactured, based on the glass composition, raw materials, melting, and way of casting, the microstructure of the glass ceramic becomes a function of the annealing step and nucleation and crystallization process. Thus, the informed adjustment and study of each step is essential to control your glass ceramic properties.

The conditions that make the preparation of a uniform glass ceramic microstructure are by no means easy. So the first step after glass preparation is to determine how to anneal the glass very well in order to produce net shape glass parts that are ready for crystallization. The way to do this is to use DTA and thermal expansion data to determine as near as possible the best temperature for the annealing. Inappropriate annealing will result in several problems in the glass and consequently in the resultant glass ceramic.

A badly annealed glass will contain very high internal residual stresses that will make it very difficult to cut and handle the glass into precise shapes. Also, it is highly likely that extensive cracks will appear on cutting. In addition, the thermal expansion behavior will be shifted from that of the best annealed glass.

Another step is to adjust and choose the best heat treatment schedule to perform proper nucleation and achieve controlled crystallization without causing deformation of the products. The most appropriate heat treatment schedule can be determined from the transition temperature, transition range, softening temperature, along with both the DTA and the thermal expansion curves.

Another step is to define the best rate of heating and the best rate of cooling to achieve uniformly microcrystalline glass ceramics. The most significant factor in this condition is to choose a proper heat treatment rate for the nucleated body up to the crystallization temperature that must be low enough to perform the crystallization without being subjected to deformation. The heating rate in fact will differ according to the nature of the crystalline phase to be developed. Under strict conditions of optimization and microstructural adjustment, the average crystal size in a glass ceramic can be controlled to be submicron and a crystal size in the range of 20–50 nm can be achieved, which corresponds closely to the crystal size obtained in nanoceramics made using the chemical sol–gel technique.

The random orientation of the crystals in the glass ceramic is a consequence of the way in which the crystals are formed by precipitation from the glass under conditions where there are no stresses that can cause alignment of the crystals. Uniform finely divided fluorophlogopite glass ceramics are shown in Figs. 8.8 and 8.9.

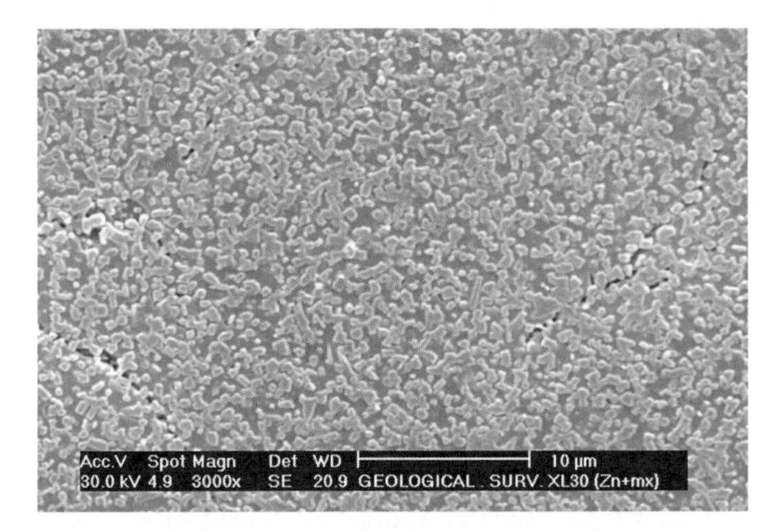

Fig. 8.8 Typical microstructure for a finely divided fluorophlogopite glass ceramic

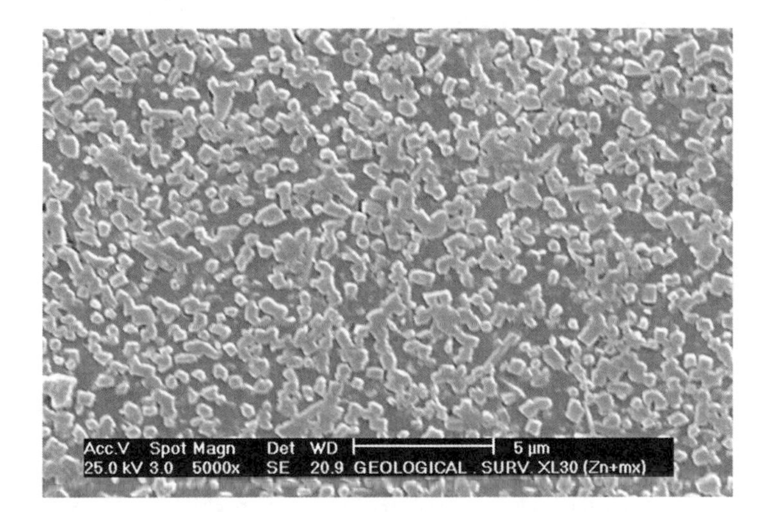

Fig. 8.9 Typical microstructure for a finely divided fluorophlogopite glass ceramics

8.4 The β-Spodumene/Fluorophlogopite System

An interesting example is the development of β-spodumene simultaneously with fluorophlogopite in a glass ceramic. The fluorophlogopite constitutes the primary mineral phase together with β-spodumene solid solution, in addition to substantial amounts of forsterite as a secondary mineral phase. Fluorophlogopite grains are uniform in size and randomly distributed in the glass matrix.

The β-spodumene phase occurs in a considerable amount in bodies designed to have between 10 and 30% of spodumene. Oxides such as MgO, MgF_2, and B_2O_3 are present in the compositions. Their presence assures the formation of fluorophlogopite phase with β-spodumene. P_2O_5 and B_2O_3 may be incorporated into the residual glassy matrix to lower the fusion temperature and alter the coefficient of thermal expansion. It has been revealed that crystal growth is superimposed upon a phase-separated glass.

The phase separation plays an influencing role in the nucleation and crystal growth rates. The platy nature of mica platelets being grouped in clusters of about 1–2 μm in diameter and grown around the phase-separated droplets are believed to precede the mica growth. Typically a uniform, well controlled and finely divided equigranular microstructure of fluorphlogopite can be achieved.

The SEM micrographs (Figs. 8.10) clearly portray the fluorophlogopite crystals, which are the large plate-like and tabular grains, and β-spodumene solid solution grains. The β-spodumene is extremely fine-grained with grains having diameters less than about 1 μm and difficult to differentiate visually. Nevertheless, the presence of β-spodumene solid solution is confirmed by XRD and DTA analyses.

The presence of β-spodumene solid solution reduces the coefficient of thermal expansion to be in the range of 36 to 31×10^{-7}/°C, such that glass ceramics in this system demonstrate excellent resistance to thermal shock. In order to achieve a coefficient of expansion of less than 40×10^{-7}/°C, β-spodumene solid solution should comprise at least about 25% of the total crystallinity as shown by bodies designed to develop 30% spodumene. Thus, the interlocking of the fluorophlogopite crystals coupled with a relatively high percentage of fine-grained β-spodumene solid solution reduces the coefficient of thermal expansion to less than 35×10^{-7}/°C. This compares with a coefficient of thermal expansion for fluorophlogopite mica of 90×10^{-7}/°C.

The crystallization of the fluorophlogopite crystals renders the glass ceramics machinable and can be used for the CAD–CAM technique. Unfortunately, the β-spodumene phase also tends to harden the body and thereby impair the machinability character according to their amount, growth, and distribution in the glassy matrix. In bodies designed to have 60% crystalline β-spodumene, the machinability is impaired due to an excessive increase in the hardness, although the material demonstrates a negative coefficient of thermal expansion (-27×10^{-7}/°C). Therefore, when the proportion of β-spodumene solid solution increases, the hardness increases sharply and the glass-ceramic is no longer considered machinable as demonstrated by a body designed to contain 60% spodumene when fired at 950°C.

The best machinability is secured at the highest concentration of fluorophlogopite and wherein the fluorophlogopite has a grain size of less than 5 μm. However, its content does not permit a reliable decrease in the expansion or further increase in the value of hardness.

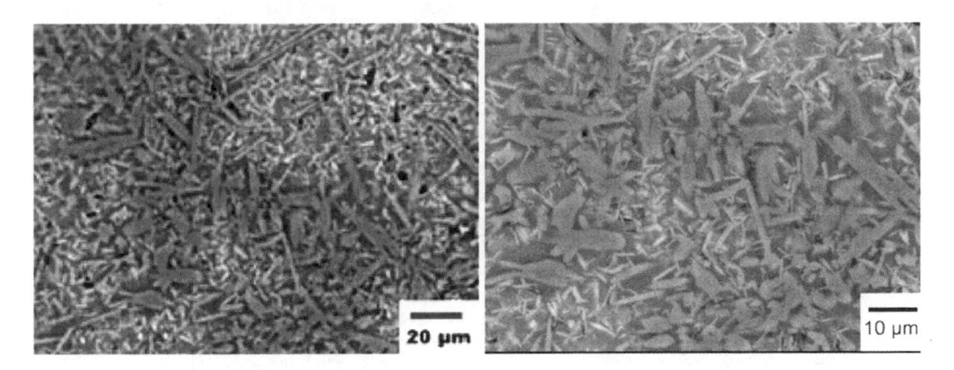

Fig. 8.10 Mica glass ceramics designed to contain 10% spodumene and prepared at 950°C showing interlocking fluorophlogopite crystals with fine-grained β-spodumene

The inclusion of nucleating agents such as TiO_2 and/or ZrO_2, which are normally utilized to nucleate β-spodumene solid solution crystals, is not necessary in the system described above. It is possible that the presence of such nucleating agents might encourage the development of parasitic phases. In contrast, the addition of B_2O_3 is an important factor for the crystallization of fluorophlogopite-based glass ceramics that exhibit good mechanical machinability. Fluorine loss from the melt definitely affects the crystallization and should be kept below 20% at the melting temperatures employed. Volatilization of fluoride from the melt can be quite low at the melting temperatures employed.

The rate of crystal growth is dependent upon the heat treatment regime. Hence, at temperatures slightly above the transition range, crystal growth is quite slow and the glass body is susceptible to deformation. However, the heating rate for the glass ceramic at temperatures above the transition range ought not to be so rapid as not to allow adequate time for appropriate grain growth. Heating rates of 10°C/min or higher can be successfully employed. The use of a two-step crystallization heat treatment is preferred, as this will minimize the possibility of deformation, since substantial nucleation ensures more rapid and uniform subsequent crystallization. A possible heat treatment that is acceptable consists of a nucleation phase at 650°C/3 h, before subsequent crystallization at temperatures between 900 and 1,050°C to ensure uniform crystallization.

The desired crystallization is attained by exposing the glass to a temperature within the crystallization range in like manner when the melt is simply cooled to room temperature. Hence, again, the occurrence of phase separation and nucleation at temperatures above the crystallization range is critical to crystallization.

Typically, the glass ceramic bodies exhibit an average coefficient of thermal expansion of less than 40×10^{-7}/°C in the temperature range of 20–700°C. The thermal expansion results of compositions based on 10, 30, and 60% spodumene content for samples 1, 2, and 3, respectively are displayed in Fig. 8.11. The transition temperatures are 510, 510, and 460°C, the softening points are 550, 550, and 520°C, and the expansion coefficients are 36.21, 31.03, and -27.60×10^{-7}/°C for 10, 30, and

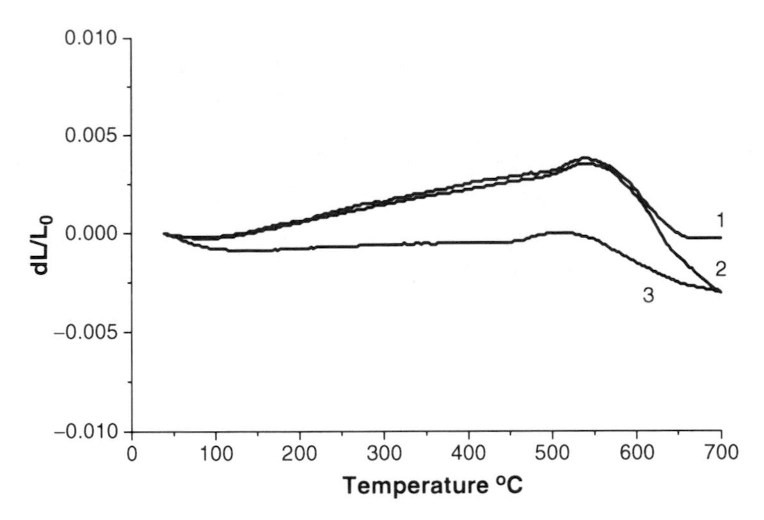

Fig. 8.11 Thermal expansion of different spodumene mica glass ceramic compositions. 10% spodumene–mica. 30% spodumene–mica. 60% spodumene–mica

60% spodumene, respectively. A distinct similarity is apparent in the expansion behavior of both compositions based on 10 and 30% spodumene–mica. The two samples showed closely matched expansion behavior.

K_2O and Na_2O have the effect of encouraging the formation fine mica crystals during the precipitation by heat treatment and improving the chemical durability. The chemical durability depends upon the amount of a residual glass in the glass ceramic. With a decrease in the amount of the residual glass, the chemical durability is considerably improved. The addition of K_2O in particular improves the chemical durability.

8.5 Leucite–Fluorophlogopite Glass Ceramics

Another example is the optimization of the microstructure of a low fusing temperature leucite–fluorophlogopite glass ceramic for dental restorations. Leucite is known to crystallize by surface crystallization mechanism, while fluorophlogopite mica can crystallize by either surface or bulk crystallization mechanism. The development and optimization of leucite–fluorophlogopite glass ceramics for coating chromium–nickel–molybdenum alloys that can mature at a low fusing temperature in as little as 2 min between 800 and 900°C is a difficult subject. Although the leucite glass ceramic fused to metal restorations account for approximately 80% of fixed restorations worldwide, their mechanical strength is poor (50–60 MPa), a fact that impairs the clinical survival rates due to chipping or delamination of the veneer. To solve this problem we need to optimize the microstructure in order to preserve compromised mechanical, thermal, and chemical properties.

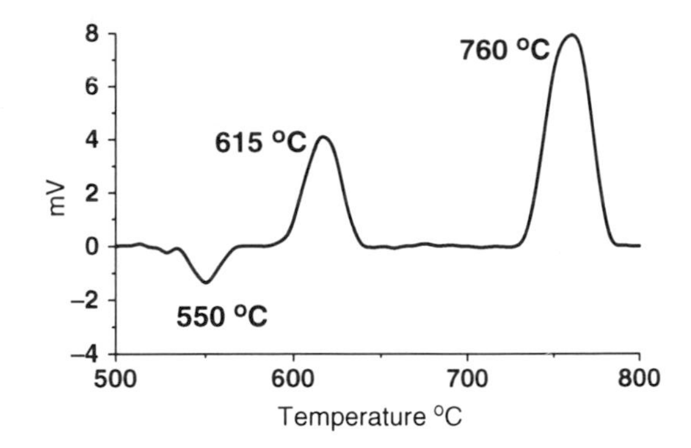

Fig. 8.12 DTA of a leucite fluorophlogopite glass ceramics

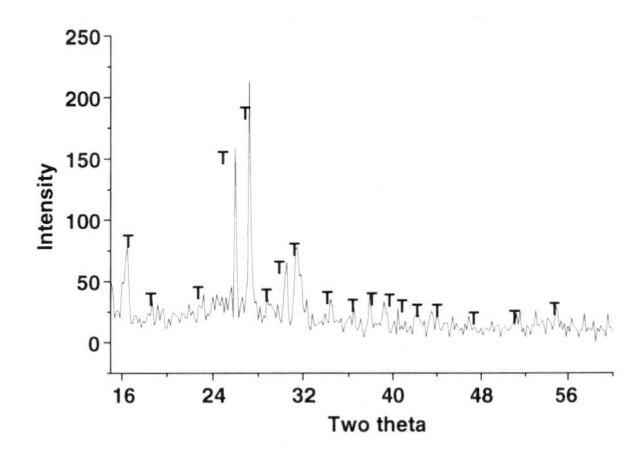

Fig. 8.13 XRD showing single phase tetragonal leucite; *T* tetragonal leucite

The optimization of the microstructure by the simultaneous crystallization with interlocking tetragonal leucite and platy fluorophlogopite is guided by the results of DTA in Fig. 8.12. The crystallization temperatures of tetragonal leucite and fluorophlogopite were determined to be 615 and 760°C, respectively, and the maturing temperature was estimated to be 870°C for 2 min.

After optimization, following the DTA results and adjustment of the chemical composition, the phase analysis by XRD is shown in Figs. 8.13 and 8.14.

A low temperature fusing leucite–fluorophlogopite glass ceramic, with enough strength (90–110 MPa) and good thermal compatibility, was achieved by the simultaneous crystallization of platy mica together with tetragonal leucite colonies. Uniform tetragonal leucite arranged in colonies with a grain size less than 2 μm is shown in Fig. 8.15. A successful microstructural optimization by the simultaneous crystallization of evenly dispersed platy grains of fluorophlogopite together with tetragonal leucite crystals is shown in Fig. 8.16. The leucite glass ceramics was reinforced by the crystallization of platy mica crystals.

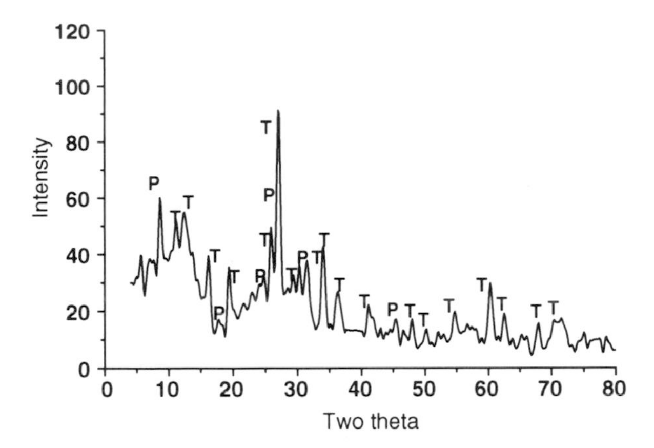

Fig. 8.14 XRD showing leucite–fluorophlogopite glass ceramics, *T* tetragonal leucite, *p* fluorophlogopite

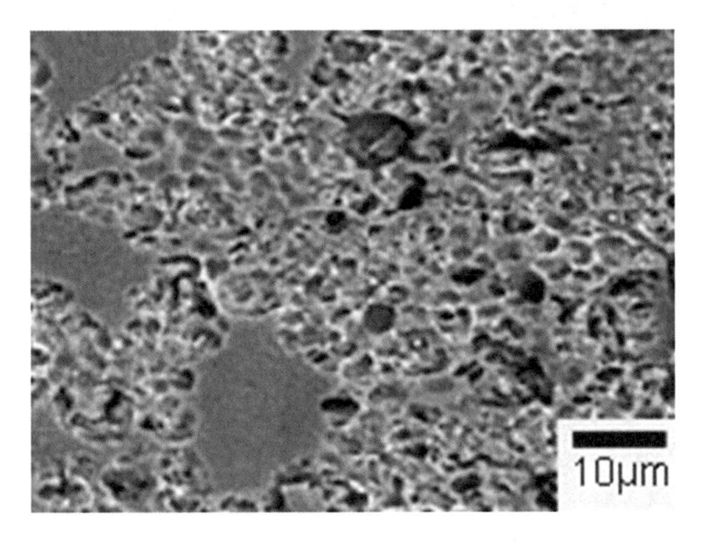

Fig. 8.15 SEM of low fusion leucite glass ceramics showing uniform tetragonal leucite colonies

The coefficient of thermal expansion is optimized to be $14.0 \pm 0.5 \times 10^{-6}$/°C, the transition temperature is 525°C, and softening temperature is 640°C, which is compatible with coating a chromium–nickel–molybdenum alloy. The microhardness and the chemical solubility were found to be 450 ± 20 HV and 30 ± 10 μg/cm^2, respectively, to resemble the properties of the natural teeth. Leucite–fluorophlogopite glass ceramics with fine uniform microstructure have been prepared and a significant improvement of mechanical strength has been realized.

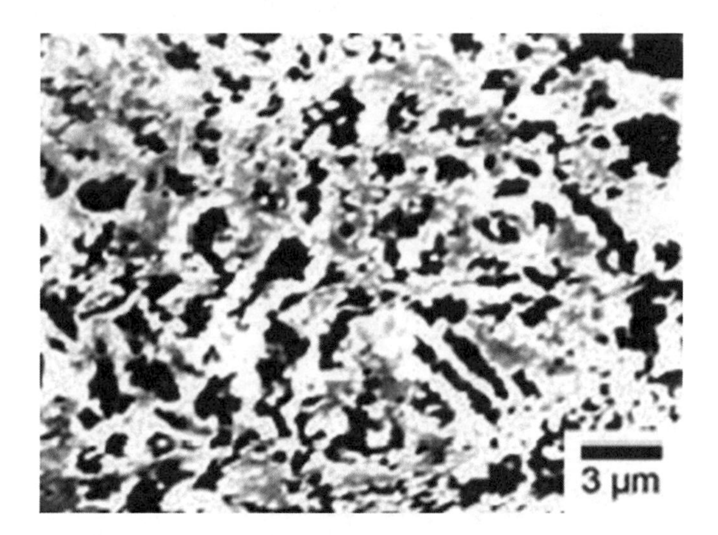

Fig. 8.16 SEM, XRD, and DTA of low fusion leucite–fluorophlogopite glass ceramics with maturing temperature 850°C/2 min

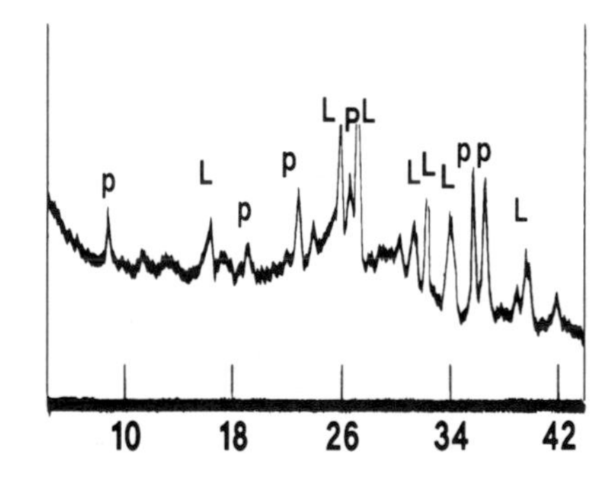

Fig. 8.17 XRD of a glass ceramic showing cubic leucite and fluorophlogopite

On the other hand, there is a possibility to crystallize cubic leucite together with fluorophlogopite mica as displayed by XRD in Fig. 8.17 and SEM in Fig. 8.18. Although the crystallization of cubic leucite improves the mechanical properties, it raises the transition temperature and the crystallization temperature, and the resultant glass ceramic is no longer a low fusion material.

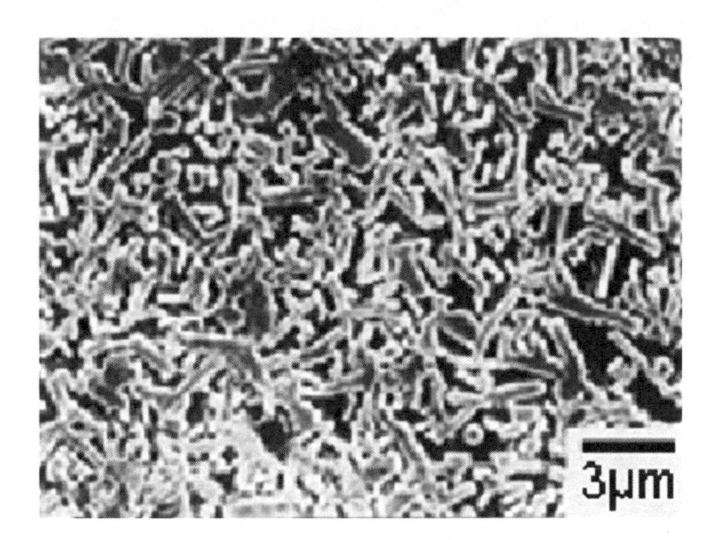

Fig. 8.18 SEM of leucite–fluorophlogopite glass ceramics, maturing temperature 950°C/1 h

8.6 Fluorcanasite Dental Glass Ceramics

Glass-ceramics based on chain silicates, e.g., enstatite ($MgSiO_3$), potassium fluorrichterite ($KNaCaMg_5Si_8{\cdot}O_{22}F_2$), and fluorcanasite ($K_2Na_4Ca_5{\cdot}Si_{12}O_{30}F_4$), are single, double, and multiple chain silicates, respectively. These glass ceramics have the potential to exhibit a combination of high fracture strength and high fracture toughness.

Canasite is a multiple chain silicate glass ceramic exhibiting an anisotropic, lath-like crystal structure with a randomly oriented interlocking bladed morphology. Structurally, the crystals are composed of parallel silicate chains crosslinked to make a long box-like backbone in which the potassium ions rest.

The typical morphology of canasite ceramics exhibits a lath-like crystal structure that grows in situ with a thickness generally less than about 1 µm, a width commonly varying between 0.25 and 2 µm, and a length normally ranging between 2–10 µm. Crystals of larger dimensions can be developed by reducing the rate of nucleation or increasing the rate of crystal growth. The microstructure of fluorocanasite is shown in Fig. 8.19. Those modifications can be achieved via changes in precursor glass composition and/or in the crystallization heat treatment.

Glass ceramics, comprising canasite as the sole crystal phase, display a modulus of rupture in excess of 355–420 MPa. A problem may arise from the crystallization of agrellite ($NaCa_2Si_4O_{10}F$) together with canasite, which extensively reduces the mechanical strength by up to 50%. This mineral phase does not demonstrate as extensive an interlocking morphology as canasite (Table 8.1).

The inclusion of Al_2O_3 and/or B_2O_3 in the parent canasite glass compositions limits the unusual crystal growth and promotes the development of fine-grained crystals, while

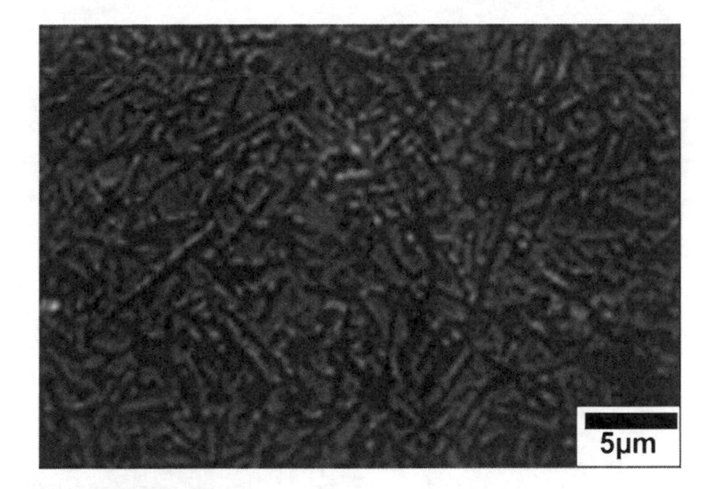

Fig. 8.19 Fluorcanasite microstructure

Table 8.1 The variation of the mechanical strength of canasite with the variation in phase composition

	Compositions (wt%)		
Oxide	1	2	3
SiO_2	57.0	55.7	60.9
CaO	12.9	13.2	19.8
CaF_2	13.0	11.1	12.0
Na_2O	9.8	9.6	7.9
K_2O	7.4	6.3	–
Al_2O_3	–	–	–
ZrO_2	–	1.2	–
BaO	–	3.1	–
T_g (°C)	530	470	
Heat treatment	700°C/2 h–900°C/4 h	700°C/2 h–900°C/4 h	700°C/2 h–875°C/4 h
Phases	Canasite	Canasite	Agrellite
MOR (MPa)	400	390	170

CaF_2 acts as the nucleating agent. White opal colors may results from the presence of CaF_2 crystallites that can be avoided by cooling the melts as rapid as possible. Also, loss of fluorine through volatilization during melting should not exceed 5–10 wt%.

Addition of ZrO_2 is advantageous in four respects as it supplements the nucleation role of CaF_2, allowing the development of fine-grained bodies and inhibits the thermal deformation of the precursor glass during the heat treatment. Zirconia also enhances the whiteness or opacity of the glass-ceramic body and improves the chemical durability of the glass-ceramic body.

Further Reading

Doherty, P.E., Lee, D.W., Avis, R.S.: Direct observation of the crystallization of Li_2O-Al_2O_3-SiO_2 glasses containing TiO_2. J.Am.Ceram.Soc. **50**(2), 77–81 (1967)

El-Meliegy, E.M.: Preparation and characterization of low fusion leucite dental porcelain. Br Ceram Trans **102**(6), 261–265 (2003)

El-Meliegy, E.M.: Machinable spodumene- fluorophlogopite glass ceramics. Ceram Int **30**(6), 1059–1065 (2004a)

El-Meliegy, E.M.: Low fusion fluorophlogopite- leucite containing porcelain. Br Ceram Trans **103**(5), 231–234 (2004b)

El-Meliegy, E.M., El-Bassyouni, G.T.: Study of the bioactivity of fluorophlogopite- whitlockite ceramics. Ceram Int **34**(6), 1527–1532 (2008)

Hamzawy, E.M.A., El-Meliegy, E.: Crystallization in the Na_2O-CaO-Al_2O_3-SiO_2-(LiF) glass compositions. Ceram Int **33**(2), 227–231 (2007)

Hamzawy, E.M.A., El-Meliegy, E.: Preparation of nepheline glass-ceramics for dental applications. Mater. Chem. Phys. **112**(2), 432–435 (2008)

Ibrahim, D.M., El-Meliegy, E.: Mica leucite dental porcelain. Br Ceram Trans **100**(6), 260–264 (2001)

Lee, W.B., Rainforth, M.W.: Ceramic microstructures; property control by processing. Chapmann & Hall, London (1994)

Lewis, M.H., Johansen, J.M., Bell, P.S.: Crystallization mechanisms in glass-ceramics. J. Am. Ceram. Soc. **62**(5–6), 278–288 (1979)

Mustafa, E.: Fluorophlogopite porcelain based on talc feldspar mixture. Ceram Int **27**(1), 9–14 (2001)

Chapter 9
Development of Colour and Fluorescence in Medical Glass Ceramics

9.1 Coloured Glasses

Coloured glasses absorb light, either broadly or sharply, in the ultraviolet, visible or in the near-infrared region of the spectrum so that the different colours are displayed. The colouring glasses should contain at least one of the colouring ions to perform their functions such as Fe, Ni, Co, V, Ce, etc. incorporated in the base glass. These types of coloured glass have been used in distinctive applications according to their transmission and absorption characteristics, for example, as optical filters such as neutral density filters, blue filters, and sunglasses (Weyl 1959).

However, the coloured glass should satisfy the performance requirements in such respective applications. For example, when coloured glass is used for sunglasses, it is not only required to have the accepted transmission characteristic with regard to visible light, but it is also desirable that it should disrupt light in the ultraviolet region, which is harmful to the eye.

On the other hand, if UV absorbing coloured ion, such as Ce cation is incorporated in glass to the extent of completely absorbing light in the ultraviolet region, light in the visible region is also mostly absorbed because of its broad absorption characteristics. So the chemical composition and additives must be chosen carefully when the application is concerned with human beings. The problem can be solved by providing a coloured glass which exhibits the desired transmittance characteristics which can be achieved by incorporating colourants working in the visible region of the spectrum and completely intercept light with a selected wavelength within or near the ultraviolet region.

E. El-Meliegy and R. van Noort, *Glasses and Glass Ceramics for Medical Applications*,
DOI 10.1007/978-1-4614-1228-1_9,

9.2 Coloured Glass Ceramics

Colours are used to colour glass and glass ceramics and commonly consist of one or several vitreous substances and admixtures of one or several pigments. Colours are usually manufactured by mixing frits with the inorganic pigments during or before the melting process (fritting) or during the pulverization process of the frits to be provided in the form of powder mixtures of colouring oxides and glass (Grossman 1995).

The colouring oxides are appreciably soluble in the glass melts to a certain extent. These colour pigment should be present in particles with average diameters of between 1 and 40 μm and can be applied to the substrate in a number of ways.

The Li_2O–A_2O_3–SiO_2 composition generally provides a highly crystallized glass-ceramic and the primary crystal phase, depending on glass composition and heat treatment, may be a transparent beta-quartz solid solution, or a beta-spodumene solid solution. It has been observed that the transparent beta-quartz glass-ceramics nucleated with TiO_2 tend to exhibit an undesirable light brown tint due to the presence of both TiO_2 and Fe_2O_3 in the parent glass composition. In an opaque, white glass-ceramic, the brown tint is effectively masked.

To decolourize or mask the undesirable tint, however, is more difficult in transparent glass-ceramics. The addition of neodymium oxide as a composition additive makes the glass-ceramic sensibly colourless. This problem is another example showing the importance of the chemical composition management during colour development either in glass and glass ceramics. These colours are widely used to impart a basically a pleasing, aesthetic colour and appearance to various artistic media such as glazes, porcelain enamels, glasses, and glass ceramics.

9.3 Colourants Based on Spinel Structure

Spinel structure compounds are well known colouring inorganic pigments that are not or only slightly dissolved during the vitrification of glass powders. Crystallographically, spinels are face-centered cubic into which metal ions can be located in either tetrahedral or octahedral coordination. Spinels comprise metal oxides in recognized groupings, usually expressed by formulas. A normal spinel crystal structure, for example, may be expressed as AB_2O_4 in which A may represent a monovalent or divalent metal ion and B represents a trivalent metal ion. The sum of the metal positive valences, eight, equals to the total negative valence of the oxygen to maintain the spinel electrically neutral.

Spinels inorganic pigments have been limited to the substitution of metal ions of similar charge in a host crystal structures. The substituted metal ions can fit into the spinel crystal structure in two ways, either in occupied or unoccupied tetrahedral site or octahedral site to make a solid solution consisting of single spinel crystal structure with one or more substituted metal ions.

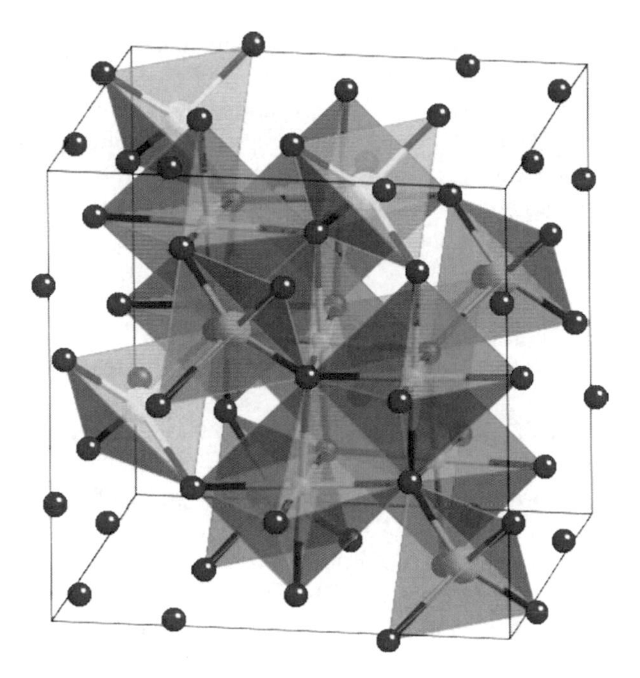

Fig. 9.1 The normal spinel is $MgAl_2O_4$ structure (O *red*, Al *blue*, Mg *yellow*; tetrahedral and octahedral coordination polyhedra are *highlighted*)

The inorganic pigment can be made of multiple spinels merged together in solid solutions. The spinel solid solution comprises a solvent spinel containing within its crystal structure the spinel-forming metal oxides; one or more spinels are mutually soluble and form a single phase relation which makes it possible to vary the colours.

Normal spinel structures (Fig. 9.1) with formula the AB_2O_4 are usually cubic closed-packed structure shared by many oxides of the transition metals in one octahedral and two tetrahedral sites that are usually used as colours. The tetrahedral sites are smaller than the octahedral sites. Trivalent cation (B^{3+}) occupies the octahedral holes, while divalent cation (A^{2+}) occupies 1/8 of the tetrahedral holes. A common example of a normal spinel is $MgAl_2O_4$.

An inverse spinel is an alternative arrangement where the divalent ions swap with half of the trivalent ions so that the divalent cation (A^{2+}) occupies octahedral sites as shown by the formula $B(AB)O_4$. A common example of an inverse spinel is Fe_3O_4.

9.4 Fluorescing Oxide Additives

Fluorescence is the emission of light by a substance that has absorbed light or other electromagnetic radiation of a different wavelength. The most striking fluorescence occurs when the object absorbs light in the UV region and emits this light in the visible region. The metameric effect results from this fact. Metamerism in the dental glass ceramics is the matching of apparent colour of objects (the proportion of total light emitted, transmitted, or reflected by a colour glass ceramic) with different light sources at every visible wavelength. Colours that match this way are called metamers. When making materials to replace natural tooth tissues, generally some types of fluorescing agents need to be added to the ceramic so as to try to produce a material that has fluorescence close to that of natural teeth.

The fluorescing glass ceramic should exhibit fluorescence based on colours that are closely related to the shades of the natural teeth. The colour of the fluorescent substance must be such as to avoid making the glass ceramic composition cloudy or reduce its clarity or translucency. Also, the ceramic must contain components that are safe to human health. The fluorescent properties of the dental ceramic must be stable at high firing temperatures and there must be no radioactivity associated with the fluorescing additives. To realize these requirements, the fluorescing materials must emit a colour that is within a greenish blue-white colour range when illuminated with ultraviolet rays at a wavelength of about 3,650 Å.

Using materials with a metameric colour match, like dental glass ceramics that should match the natural tooth, is a significant problem an industry where colour tolerances are important. The dental glass ceramics may be manufactured to provide a good colour match under a standard light source, but the match can disappear under different light sources resulting in the metameric failure.

9.4.1 Uranium Oxides

Over the years rare earths have been identified and used in soda lime glasses to produce a fluorescence or luminescence in glasses in the visible light range. The combination of uranium oxides and cerium oxides along with other rare earth oxides produces fluorescence close to that of the natural teeth. In the early days of dental ceramics various uranium compounds were used in small amounts in order to produce the desired fluorescence. The substances of the uranium family were adopted as they generally withstood the high firing temperatures used in sintering dental ceramics and exhibit fluorescence closely related to the natural teeth. Unfortunately, uranium substances are radioactive materials, and thus it is undesirable to add these to any ceramics to be used in the human body, regardless of its minute content. Moreover, although glass ceramic teeth containing uranium substances exhibit a natural fluorescence, a darkish tint in colour may appear due to the black appearance of uranium oxide.

9.4.2 Cerium and Terbium Oxides

Other fluorescing compositions that can be used as an alternative to uranium compounds are those based on cerium and terbium oxides. When these are incorporated in dental ceramics in specific amounts they are able to produce the desired fluorescence in the absence of uranium or other radioactive compounds. The use of a combination of cerium and terbium oxides with other rare earths produces fluorescence in the visible light range. In addition, the cerium–terbium oxide mixture causes glass ceramics to fluoresce under ultraviolet light in a way that is far superior to that produced by the uranium containing oxides.

While a cerium salt produces broader absorption spectra than most of the rare earth oxides, terbium oxide fluoresces in the blue-green region. At first glance this may appear undesirable in producing the wrong colour to be emitted. Fortunately, the combination of no more than 5 wt% cerium oxide and 2.5 wt% terbium oxide in glass ceramics produces a fluorescence at a somewhat higher wavelength matching that of the natural teeth (Smyth and Lee 1979).

The terbium–cerium oxides fluorescing compositions may be used with any dental glass ceramic to produce the desired fluorescence when fired to the desired temperature. The fluorescing composition may be added to the batch of the starting glass or during the processing of dental glass ceramics. The addition of cerium oxide to the frit causes bubbling and makes the frit porous as well. However, because a small amount of cerium oxide is used, there is no problem in adding the cerium oxide to the fused mass.

The amount of terbium and cerium oxides should be tested for different glass ceramic compositions in order to be able to produce the required fluorescence and avoid the adverse colouring and cost effects. In addition, the wrong content of oxide additives may complicate the process for incorporating the cerium and terbium in the glass ceramics without improving the fluorescing properties.

Salts of cerium such as the acetate, acetate hydrate, carbonate, citrate, hydroxide, nitrate, oxalate, or sulphate of cerium along with the hydroxide, nitrate, sulphate, or oxalate of terbium may be used in order to obtain the desired natural colours. The amount of such salts must be adjusted to give equivalent amounts of cerium and/or terbium oxides in the glass ceramic.

9.4.3 Europium Compounds

A dental glass ceramic, which includes an aluminium silicate, alkali, or alkaline earth aluminosilicate or an alkali or alkaline earth aluminate, can exhibit fluorescence through the addition of the rare earth metal oxide, europium. The glass ceramic fluoresces by excitation in the near ultraviolet range and shows a colour tone similar to that of natural teeth. The use of europium activator is suitable for producing fluorescence in feldspathic glasses and leucite, alumina or nepheline glass ceramics. These ceramics include the base aluminosilicate composition on which the europium depends.

It is known that the fluorescent substances containing europium as activator exhibit red fluorescence in case of trivalent europium and blue fluorescence in the case of divalent europium, which is a serious shortcoming. It has been shown in glass ceramic compositions, which contain only europium as the activator, these exhibit blue–white or violet–white fluorescence after firing. Therefore, it is supposed that the fluorescent substance is not completely decomposed by firing and most of the europium is left in its divalent state. Even if a part of europium is transformed into trivalent europium, it tends to extend its wavelength distribution to bring the fluorescent colour from blue to white, thereby contributing to a colour, which is not closely matching the natural teeth colours. A proposed solution of the problem is to substitute a part of europium by another rare earth element to extend the wavelength distribution and enable creation of fluorescent colours that are more closely related to the natural teeth.

The elements that could be used as fluorescent substances to aid the activation with europium include cerium and ytterbium as auxiliary activators, in the form of CeO_2 and Yb_2O_3, respectively, for good fluorescence with respect to the basic material. About 1 wt% of the fluorescent substances is added to the base material to obtain fluorescent dental ceramic compositions. When suitable contents of both Eu_2O_3 and Yb_2O_3 are added to a powdered dental ceramic, a glass ceramic with milky white–blue fluorescence is achieved.

Alkaline-earth oxides containing aluminosilicate glass ceramic activated with other elements such as ytterbium are not sufficiently bright or fluorescent by themselves. However, intermixing the fluorescent substances with europium improves the brightness of fluorescence and produces a superior product to those activated with europium alone.

9.5 The Colour Evaluation

When a light falls on an object, a portion of the incident light is reflected, another portion of the incident light is absorbed, and the remaining portion is transmitted. Objects are said to be opaque when the transmitted component is so small as to be negligible. Objects are said to be transparent when the reflected component is so small as to be negligible, even though the transparent object as glass still reflects up to 4% of light from the object–air interface and absorbs around 1% in its ordinary thickness.

If material is inserted in the collimated beam of light, scattering of light takes place within the material. The value of transmittance at each wavelength can be obtained by all the light that emerges from the material, regardless of each direction. Unless the material is almost perfectly homogeneous, it is usually more significant to determine its diffuse transmittance.

It is important to mention that the range of UV spectrum wavelengths is 400–10 nm, while the range of visible light spectrum wavelengths (Fig. 9.2) is 400–700 nm. The wavelengths are commonly expressed in nanometers ($1\ nm = 10^{-9}\ m$).

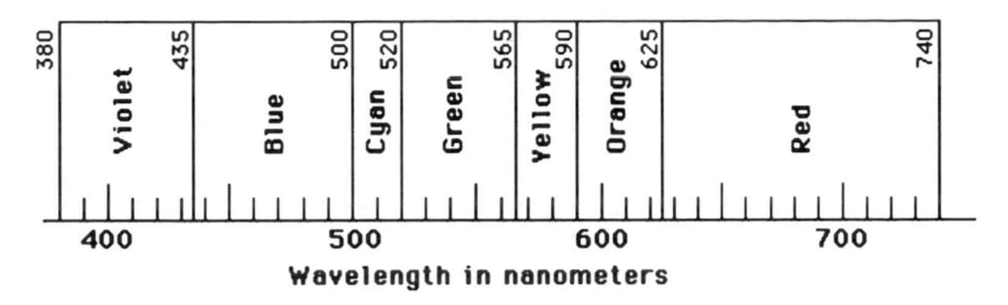

Fig. 9.2 The visible spectrum

The ability to merge a glass ceramic crown with its natural equivalent involves consideration of size, shape, surface texture, translucency, and colour. The popularity of metal–ceramic restorations is due to predictable strength achieved with reasonable aesthetics. The drawback of such restorations is the increased light reflectivity because of the opaque glass ceramic needed to mask the metal substrate.

The translucency of all-ceramic core material is one of the most important factors in controlling aesthetics and considered as a critical issue in choosing of materials. However, an increase in crystalline content to achieve greater strength generally results in greater opacity.

The translucency of glass ceramic is largely dependent on light scattering. If the majority of light passing through a ceramic is intensely scattered and diffusely reflected, the material will appear opaque. If only part of the light is scattered and most is diffusely transmitted, the material will appear translucent. The amount of light that is absorbed, reflected, and transmitted depends on the amount of crystals within the core matrix, their chemistry, and their sizes compared to the incident light wavelength.

Particles similar in size to the light wavelength have the greatest scattering effect. Both the chemical nature of the particles (leading to absorption) and the relative refractive index of the particles to the matrix affect the amount of scattering.

Material composed of small particles (approximately 0.1 μm in diameter) is less opaque when visible light passes through with less refraction and absorption in spite of the greater scattering from an increased number of particles. Large particles (approximately 10 μm in diameter) cause surface reflection as light strikes, refraction as light passes through, and absorption.

However, large-particle materials have reduced numbers of particles per unit volume and consequently display less scattering and decreased opacity. For maximal scattering and enhanced opacity, a dispersed particle slightly greater in size than the wavelength of light, and with a different refractive index to the matrix, is required. This effect is seen with zirconium oxide, which has a maximal effect of opaqueness.

9.6 Measurement of Colour

The perceived colour of an object depends on a number of factors including spectrum of light emitted form the light source, light being electromagnetic radiation in the wavelength range of 380–780 nm, the optical properties of the object, such as reflection, scattering, and absorption of the light, and the observer.

In 1905, the American artist A.H. Munsell came up with a method for describing colours, which he classified according to their hue, chroma, and value:

- *Hue*. This represents the dominant colour (i.e. wavelength) of the spectrum of light from the source. The possible colours are violet, indigo, blue, green, yellow, orange, and red. The three primary colours, from which all other colours can be produced, are red, green, and blue.
- *Chroma*. This is the *strength* of hue, in other words how vivid the colour is.
- *Value*. the brightness or darkness of the object that ranges from black to white for diffusive or reflective objects, and from black to translucent objects.

Whereas hue and chroma are properties of the object, the value will depend on the incident light, the surface finish of the object, and the background if the material transmits light.

Whereas dentists and dental technicians use shade guides to communicate colour, a more quantitative way of defining a colour is to consider the *Yxy* space, which expresses colour in terms of x and y chromaticity coordinates. A more widely used method is the *XYZ* space, which allows colours to be expressed as percentages based on the three tristimulus values X, Y, and Z.

1. *Yxy* space (Fig. 9.3) expresses the *XYZ* values in terms of x and y chromaticity coordinates, where the *XYZ* space allows colours to be expressed as percentages based on the three tristimulus values X, Y, and Z that comes from the colour perception of the eye and Y correlates to the apparent lightness.
2. The coordinates are shown in the following formulas, used to convert *XYZ* into *Yxy*:

$$Y = Y$$

$$x = X / (X + Y + Z)$$

$$y = Y / (X + Y + Z)$$

9.6.1 Quantitative Measurement of Translucency or Opacity

A quantitative measurement of translucency is determined by comparing reflectance of light (ratio of the intensity of reflected radiant light to that of the incident radiant light) through a specimen tested over an opaque black support with a low reflectance and a white support with a high reflectance as shown in Fig. 9.4. The Y value

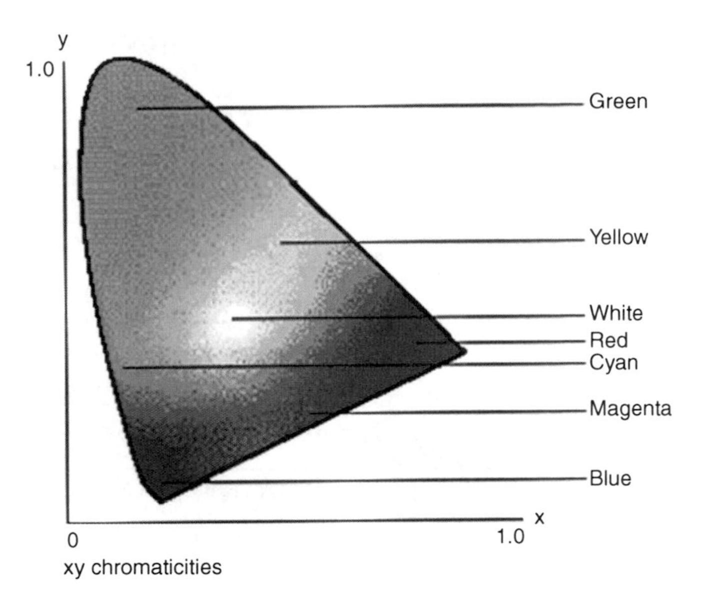

Fig. 9.3 *Yxy* space

in *Yxy* colour space represents the Reflectance, where *x* is the value of hue and *y* is the value of chroma.

The contrast ratio (CR)=Yb/Yw is the ratio of the reflectance of light of the glass ceramic on a black surface (*Y*b) to the reflectance on a white surface (*Y*w) as shown in Fig. 9.4. This ratio tends towards unity for opaque materials and towards zero for transparent materials. An integrating sphere can be attached to a calibrated spectrophotometer for specimen measurement. The specimens are measured with 0° illumination and diffuse viewing geometry. Each specimen is placed at the reflectance specimen port, and five measurements are made with the white reference support (*Y*w) and then the black support (*Y*b), resulting in a total of ten measurements per specimen.

9.6.2 The Masking Ability of Veneering Ceramics

The masking ability of a specimen can be evaluated by calculating the colour difference of the veneers over white and black backgrounds using this equation:

$$\Delta E^* = [(L^*_1 - L^*_0)^2 + (a^*_1 - a^*_0)^2] + (b^*_1 - b^*_0)^2]^{1/2}$$

where L^*_1, a^*_1, and b^*_1 represent the colour coordinates of the specimen over the white background, and L^*_0, a^*_0, and b^*_0 represent the colour coordinates of the specimen over the black background.

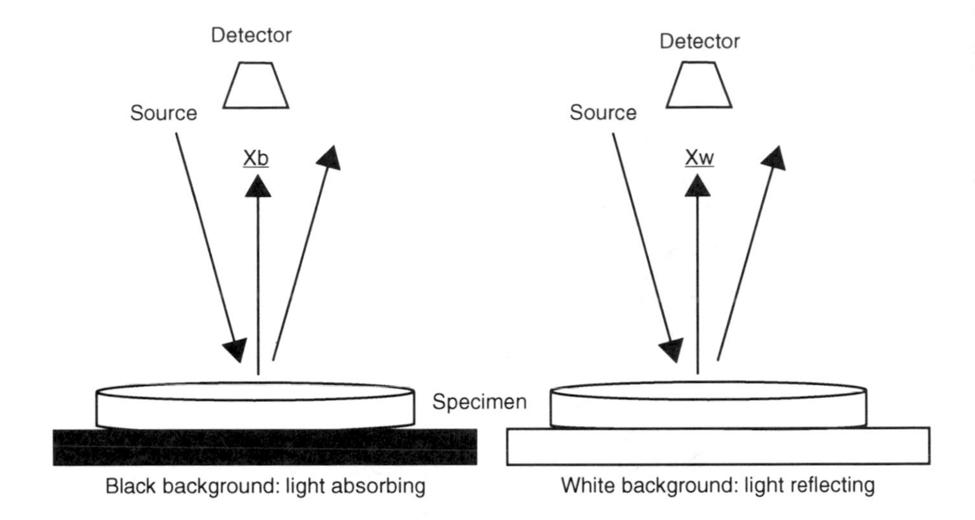

Fig. 9.4 Contrast ratio measurement

In $L^*a^*b^*$ colour space, L^* is represented by brightness (white-black), a^* for redness to greenness, and b^* for yellowness to blueness (Fig. 9.5).

The masking ability of a specimen is evaluated by calculating the colour difference of the veneers over white and black backgrounds using ΔE^*. The threshold value proposed for $\Delta E^*=5$ corresponding to CR = 0.91. Thus, none of the veneers was able to mask the colour of a black background. Clinically, the discolouration of a tooth substrate or a metal post is not as severe as the black background, therefore, ΔE^* values of the veneers should, therefore, be smaller.

9.6.3 Metamerism

An example where a colour difference measurement is useful is in relation to metamerism. The human eye contains three colour receptors (cone cells), which means that all colours are reduced to three sensory quantities, called the tristimulus values. Metamerism occurs because each type of cone responds to the cumulative energy from a broad range of wavelengths, so that different combinations of light across all wavelengths can produce an equivalent receptor response and the same tristimulus values or colour sensation.

Metamerism occurs when two objects, with different light-reflecting properties and identical colour appearance, appear different with changing the observation conditions (for example, change of the light sources).

To determine the metamerism between two objects, the $L^*a^*b^*$ values of two specimens of the same thickness (typically 2 mm) is measured relative to two standard illuminants (e.g. D65 and A). The colour differences (ΔE^* Lab) relative to the two illuminants can then be calculated.

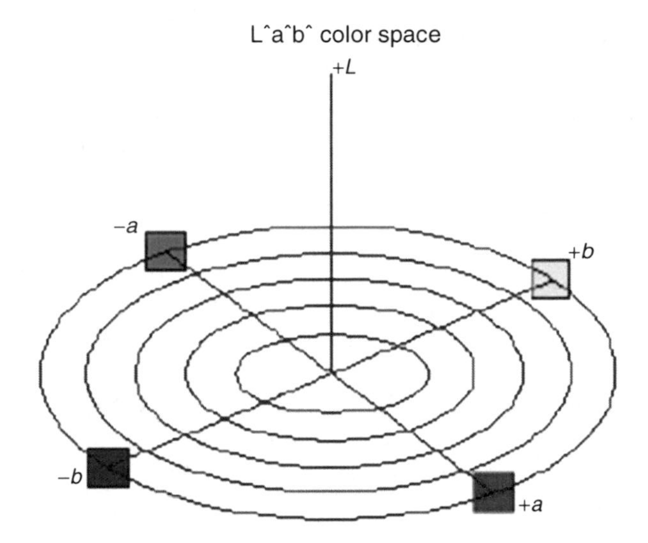

Fig. 9.5 $L^*a^*b^*$ colour space

The modified metamerism index (MI) is the ratios of colour difference due to the change of illuminant. Using Illuminant A as a reference, the colour differences (ΔE^*) between the mean CIE L^*, a^*, and b^* values for each specimen is then calculated. If there were no metameric effect between the specimens, the MI value would be 1.

9.7 Fluorescing Glass Ceramics

Natural teeth are known to have a white fluorescence under ultraviolet light. When making artificial teeth of glass ceramics, generally fluorescing agents need to be added to produce porcelains with fluorescence close to that of the natural teeth.

There are some conditions that need to be fulfilled to produce suitable fluorescing compositions, able to match that of the natural tooth. The fluorescing glass ceramic should exhibit fluorescence and natural colours having tones closely related to the shades of the natural teeth. The colour of the fluorescent substance must be white to avoid making the glass ceramic composition muddy or cloudy or reduce its clarity or translucency. Also the artificial teeth must contain components that are safe to human health.

The fluorescent glass ceramic must be proof against firing in conserving its fluorescent property and has no radioactivity. To realize these requirements, the fluorescing materials must induce the colour of the synthetic ceramic tooth to exhibit fluorescence with greenish blue–white colour range when illuminated with ultraviolet rays at a wavelength of about 3,650 Å.

9.8 Development of Colours and Florescence in UV Regions

In dental ceramics, a ceramic pigment or combination of several pigments is incorporated into the composition. The crystals of the pigments remain sufficiently intact during the firing process to impart the desired colour in the finished product. Ceramic pigments mostly consist of refractory oxides, where cations of rare earth or transition metals as Co, Cr, Fe, Mn, Ni, and V act as colour centres. The colour provided by those pigments, however, is governed to a large extent by the crystalline structure of the host compound into which the transition metal ions are incorporated.

It is quite common for the colour produced in the parent glass to be different from the colour developed in the glass ceramic crystallized from the same glass. This change is brought about through the alteration of the residual glass from which the components comprising the crystals are separated. The change also depends on the nature of the developed crystalline phase or phases and the ease of use of the colouring elements into the new crystal structure. Also the colour change may be related to the type of cations substitution in the crystal structure either partially or fully solid solution or interstitially incorporated in the structure.

The incorporation of colouring oxides such as CeO_2, CoO, CuO, Fe_2O_3, MnO_2, and TiO_2 into the glass-ceramic would likewise alter the surrounding field to produce intrinsically coloured mica glass-ceramic. The colour shade produced by CeO_2 can also be modified through the inclusion of relatively inert colourants such as Er_2O_3 for red, MnO_2 or Cr_2O_3 for gray, and TiO_2 for yellow to white (Table 9.1).

9.9 Metamerism in Glass Ceramics: The Problem and Solution

The colour of dental glass ceramic is measured in the visible light region between 380–700 nm ranges and gives perfect colour shade matching. When the synthetic dental glass ceramic is applied in the patient mouth, the ceramic colour due to changes in the lighting conditions such that the dental ceramic no longer matches the natural tooth colours. The reason may be that the artificial teeth do not fluoresce with the same colour spectrum when the illuminant is not daylight but for example ultraviolet light. Thus the artificial teeth do not behave the same as the natural teeth under different lighting conditions. This metameric effect is a complex issue to the patient as his/her teeth appear different in colour under different illuminations.

Therefore, the artificial teeth must contain fluorescing components to induce the natural fluorescence in the glass ceramic. The fluorescing composition must be safe to human health within secure criteria and proof against firing temperature in conserving its fluorescence characteristics and not radioactive. To realize these requirements, the fluorescing components ideally should induce the glass ceramic tooth to exhibit fluorescence with greenish blue–white colour when illuminated with ultraviolet ray at wavelength of about 3,650 Å.

Table 9.1 Different colour shades by different oxides addition

Colourants	Wt%	Glass	Glass-ceramic
CoO	<0.5	Dark purple blue	Purple blue
CuO	<2	Cyan	Gray
Fe_2O_3	<1	Gray green	Gray
MnO_2	<2.5	Purple	Pale purple
$Fe_2O_3+MnO_2$	1+2.5	Brown	Amber
$Fe_2O_3+MnO_3$	0.2+2	Pale brown	Pale amber
CeO_2+TiO_2	5+5	Brown	Ivory
CeO_2+TiO_3	0.2+5	Pale yellow	White
CeO_2	2–5	Red brown	Ivory

There are some conditions that need to be fulfilled to produce suitable fluorescing compositions, able to match that of the natural teeth. The fluorescing composition should exhibit fluorescence and natural colours having tones closely related to the natural teeth shades, i.e. the colour of the teeth should be the same in visible light and UV illumination. The colour of the fluorescent substance must be white, so it does not make the glass ceramic cloudy or reduce its clarity or translucency.

When making artificial glass ceramic teeth, some types of fluorescing agents are added to produce porcelains with fluorescence close to natural teeth. In order to produce acceptable colours for use in dental glass-ceramics, the colourants to be added should produce yellow to reddish yellow glass ceramics. Except uranium, no single transition metal or rare earth metal cation has been identified to yield a clear yellow effect. However, the regulations regarding radioactivity restrict the use of uranium. However, using a combination of Ce and Ti oxides can give yellow colours in glazes and glasses. The use of a combination of CeO_2 and V_2O_5 in soda lime silica glass also produced a very pale yellow colour.

The use of a combination of CeO_2 and V_2O_5 also produces a very pale yellow colour. The use of CeO_2 itself produced no colour in glass, but in combination with TiO_2 imparted strong yellow colours. In the case of crystallization of the glass into glass ceramic it gives a yellow colour as shown in Table 9.2. The yellow colour can be modified by incorporating the rare earth metal oxide Er_2O_3. The combination of CeO_2 and TiO_2 works well in the visible light, but fluorescence under long wave ultraviolet illumination does not happen, because the presence of TiO_2 strongly absorbs ultraviolet radiation resulting in the problem of colour change with light (Ying et al. 2002).

Other fluorescing compositions based on cerium and terbium oxides produce fluorescence under ultraviolet light. It is known that the concentration of terbium oxide in vitreous porcelain fluoresces at a somewhat higher wavelength in the blue-green regions. The fluorescence of terbium oxide combined with the absorption spectra for cerium oxide surprisingly produces fluorescence similar to that of the natural teeth.

In the visible light, CeO_2 is the key ingredient for yellow colouration. CeO_2 also acts as a strong fluorescing agent in glass-ceramic and imparts a yellow fluorescence under long wavelength (3,650 Å) ultraviolet light. A glass ceramic of

Table 9.2 Different colours produced in mica glass ceramic by CeO_2

Nucleation		Crystallization		Glass-ceramic colour
°C	Time (h)	°C	Time (h)	CeO_2 (2–5) wt%
650	1	1,000	4	Yellow
650	0.5	1,050	2	Glassy
650	0.5	1,050	3	Glassy
650	1	1,050	3	Yellow
650	0.5	1,100	3	Glassy
650	1	1,100	3	Yellow
650	0.5	1,150	3	Yellow opaque

alkali or alkaline earth aluminosilicate can exhibit fluorescence through the addition of an activator such as europium. The glass ceramic is fluorescing by excitation with near ultraviolet ray and show a colour tone similar to that of natural teeth.

The use of europium activator is suitable for producing fluorescence and aids in avoiding dull colours in leucite/feldspar glass ceramics, which include the base aluminosilicate composition in which the europium is loaded contains SiO2, Al2O3, K2O, B2O3, Na2O, BaO, and MgO. The dental glass ceramic usually includes fine crystalline phases dispersed in large quantity of glassy phase and matures when fired at 900–1,000°C.

The fluorescent substances containing europium as activator exhibit red fluorescence in case of trivalent europium and the blue fluorescence in case of divalent europium. The porcelain compositions, which contain only europium as the activator, exhibit undesirable blue–white or violet–white fluorescence after firing. A proposed solution of the problem of colour mismatch is to substitute a part of europium by another element, whereby the wavelength distribution is extended to enable creation of fluorescent colours, which are more closely related to the natural teeth.

9.10 Opalescence

What is commonly referred to as the "vitality" of a tooth is the combined effect of colour, translucency, fluorescence, and opalescence. Opalescence, which is also known as the "opal effect", is a light scattering phenomenon of the shorter wavelengths of the visible spectrum in translucent materials, and it produces a bluish appearance when viewed with reflected light and an orange/brown appearance when viewed via transmitted light. The definition of opalescence is the visual property of something having a milky brightness and a play of colours from the surface.

The enamel of natural tooth is opalescent; therefore, considering the increasing demand for aesthetic restorations, dental restorative materials should reproduce the opalescence of natural tooth. Restorations that are not opalescent cannot simulate the natural "opal effect" present in natural teeth, which is particularly evident at the incisal edge of the tooth enamel.

There are three kinds of glass that are known to be opalescent. One is blue-tinged, semi-opaque or clear glass with milky opalescence in the centre. The colour is produced by the slow cooling of the molten glass in those parts that are thick causing some crystallization inside the glass. This contemporary opalescent glass was first produced in the 1920s and 1930s by companies in France such as Lalique, Sabino, and Jobling.

The second kind of opalescent glass is hand-blown and is normally made from two layers of glass by mixing different colours of glass together while hot. Strictly speaking, this is more appropriately described as iridescence. The definition of a material being iridescent is that it is varying in colour when seen in different lights or from different angles, such as the wings of a dragonfly, butterfly and beetles and mother-of-pearl, with a fluid molten look to it.

Iridescent glass has a lustrous sheen, similar in appearance to mother-of-pearl or the wings of some butterflies and beetles, with a fluid, molten look to it. The glass is made up of various glasses of different colours being mixed together while still hot, allowing them to partially blend and form the multicoloured swirls and whorls characteristic of the style. Favrile glass is a type of iridescent art glass designed by Louis Comfort Tiffany. It was patented in 1894 and first produced in 1896. It differs from most iridescent glasses because the colour is ingrained in the glass itself, as well as having distinctive colouring.

The third kind of opalescent glass has a milky white edge or a white raised pattern decorating a coloured pressed glass item. The effect is produced by re-heating the parts of the molten glass just as it has started to cool. Heat-sensitive chemicals in the glass turn the re-heated sections white.

The degree of opalescence of dental restorative materials can be determined from the opalescence parameter (OP), which refers to the difference in yellow–blue colour coordinate (CIE Δb^*) and red–green colour coordinate (CIE Δa^*) between reflected and transmitted colours, according to the following expression:

$$\mathrm{OP} = \left[(\mathrm{CIE}a^*_{\mathrm{T}} - \mathrm{CIE}a^*_{\mathrm{R}})^2 + (\mathrm{CIE}b^*_{\mathrm{T}} - \mathrm{CIE}b^*_{\mathrm{R}})^2 \right]^{1/2}$$

The subscripts T and R indicate the transmitted and reflected colour, respectively. The OP value for bovine enamel has been reported to be in the range of 7.6–22.7 and that for human enamel is 19.8–27.6 (Lee and Yu 2007).

The presence of dispersed particles or a phase-separated glass will create opalescence in dental ceramics. Opalescence can also be created through the introduction of TiO_2 or bone ash. By adding bone ash to a glass, upon re-heating, the piece will turn a pearly white colour. As yet little work has been done on the opalescence of dental ceramics and what has been done suggests that opalescence is virtually non-existent in the opaque high strength core ceramics and is at best inadequate in the veneering ceramics (Cho et al. 2009).

References

Lee, Y.K., Yu, B.: Measurement of opalescence of tooth enamel. J. Dent. **35**, 690–694 (2007)
Cho, M.S., Yu, B., Lee, Y.K.: Opalescence of all-ceramic core and veneer materials. Dent. Mater. **2**(5), 695–702 (2009)
Ying, S., Wang, Z., Tian, J., Cao, X.: Coloration of mica glass-ceramic for use in dental CAD/CAM system. Mater. Lett. **57**, 425–428 (2002)
Smyth, M.B., Lee, Y.J.: Fluorescent artificial teeth. US Patent: 4170823 (1979)
Grossman, D.G.: Colored glass-ceramic articles. US Patent: 5387558 (1995)
Weyl, W.A.: Colored glasses. Dawson's of Pall Mall, London, England (1959)

Part IV
Models of Dentally Used Glass Ceramics

Chapter 10
Leucite Glass-Ceramics

The key variables in the design of a medical glass-ceramic are glass composition, phase assemblage, and the crystalline microstructure, which are the driving factor for the different properties. The phase assemblage (the types of crystals and the proportion of crystals to glass in the glass-ceramic) is responsible for almost all of the physical and chemical properties, including the functional, microstructural, thermal, mechanical, and chemical characteristics.

The current part deals with glass-ceramics for dental applications including leucite veneering ceramics, mica glass-ceramics, and lithium disilicate glass-ceramics. Leucite is a widely used glass-ceramic for veneering metal alloys and all ceramic cores. Metal alloys include gold alloys, palladium alloys, and chromium–nickel–molybdenum alloys. Lithium disilicate and mica are used for making glass-ceramic cores and processed using CAD/CAM technology. This chapter also deals with the problems encountered during the manufacturing, processing and characterization of leucite, mica and lithium disilicate glass-ceramics.

There are essentially two ways in which leucite glass-ceramics are used in the dental field, either as feldspathic veneering ceramics for coating different metal and ceramic substructures or as a leucite reinforced glass-ceramic for resin bonded veneers, crowns and bridges. Leucite glass-ceramics are extremely adaptable and can be processed into dental prostheses via many mechanisms including, traditional sintering, hot pressing and computer aided manufacture.

The main concerns when designing glass-ceramics veneers to metal or ceramic substructures are that (1) they must adhere well to the underlying metal substructure, which requires them to have an appropriately matched coefficient of thermal expansion to handle the contraction during the cooling and (2) they have the necessary aesthetics to produce the appropriate tooth shade and be opaque enough to mask out the metal color. The controlled crystallization of the tetragonal leucite mineral phase enhances the thermal compatibility for bonding to metallic or ceramic substructures. The typical volume fraction of crystalline leucite in the glass-ceramic for veneering purposes varies between 5 and 40% leucite in the glassy matrix

E. El-Meliegy and R. van Noort, *Glasses and Glass Ceramics for Medical Applications*,
DOI 10.1007/978-1-4614-1228-1_10, © Springer Science+Business Media, LLC 2012

When the leucite is used as reinforcing fillers in resin bonded all ceramic restorations the primary requirement is that the glass-ceramic can be directly bonded to enamel and/or dentine. Irrespective of whether the leucite is processed by the sintering of powders, hot pressing, or constructed from CAD/CAM blocks, the second most important requirement is that the material must have enough strength and toughness. In these applications the coefficient of thermal expansion is not a major concern and most leucite content can be much higher as long as the aesthetics are not compromised.

This chapter deals with the different ways of adjusting the chemical composition and heat treatment schedules to get low temperature fusing leucite glass-ceramics. Various examples of leucite dental ceramics, along with their chemical compositions, thermal properties, mechanical properties, and microstructure, are discussed. We also explore how it is possible to control the microstructure of the leucite dispersed in the glassy matrix, especially in compositions with a high alkali content and low alumina content. In addition, we discuss how one might control ceramic properties such as mechanical strength, microhardness, chemical resistance, and aesthetics by the addition of various mineralizers, colorants, opacifiers, and fluorescing agents.

10.1 Industrial Importance of Synthetic Leucite

The properties of synthetic leucite can be readily modified and adjusted by careful management of the starting glass chemical composition. The microstructure, thermal behavior, mechanical properties, and optical properties are all subject to being controlled through effective modification of the composition and heat treatment schedules.

The optical properties of the leucite glass-ceramic make it one of the most appropriate materials for the fabrication of dental restorations. A major benefit in using glass-ceramics containing the tetragonal leucite phase is that the leucite phase has substantially the same refractive index as the glass, and thus, the translucency is never hindered by the crystallization of the leucite in the glass. This fact contributes to the excellent translucency of the leucite glass-ceramic. Add to this the ability of the leucite glass-ceramic to match the color of the natural tooth and what one gets are outstanding aesthetic properties.

The leucite phase provides another distinct advantage in that it is able to retain its crystalline structure at the fusion temperature. Thus the properties are not affected when the glass-ceramic is subjected to a veneering firing cycle. Also, it is believed that leucite crystallites in the glass matrix serve as nuclei during the fusion and cooling processes. The stability of leucite crystallites during fusion makes it easier to be applied to the metal or ceramic substrate without impairing the thermal expansion coefficient or the mechanical strength.

Another important feature of the leucite glass-ceramic is the simplicity with which it is possible to modify its thermal expansion coefficient through compositional variation. The modification is made by controlling the crystalline content as well as the glassy phase matrix composition to reach the required thermal expansion coefficient. Since leucite has two crystalline phases, including the cubic leucite phase with a thermal expansion coefficient of $\sim 30 \times 10^{-7}/°C$ and the tetragonal leucite phase with thermal expansion coefficient exceeding $220 \times 10^{-7}/°C$, it is possible to manipulate the coefficient of thermal expansion of the final glass-ceramic by adjusting the relative amounts of these two phases.

10.2 The K_2O–Al_2O_3–SiO_2 Phase Diagram and Related Systems

The K_2O–Al_2O_3–SiO_2 phase system, which crystallizes the leucite phases, is ideally suited as a glass base composition for dental applications. The leucite ceramic, containing leucite crystallites with a particle size of less than about 10 μm, is successfully used in the fabrication of dental restorations. Leucite crystallizes, leaving an amorphous matrix with a different chemical composition from that of the starting glass. The leucite crystallizes by the heat treatment of a precursor glass containing K_2O, Al_2O_3 and SiO_2, together with other components such as alkali fluxes, nucleating agents, and grain growth inhibitors.

The leucite can also be obtained by incongruent melting of potash feldspar to provide the required leucite content in the glassy matrix. To crystallize a significant percentage of leucite from potassium aluminosilicate feldspar, a minimum content of about 12 wt% K_2O is required. Additional alkali oxides in feldspars can serve certain functions such as Na_2O fluxing agent to reduce the melting temperature and CaO and MgO that can improve the flow during casting.

The crystallization of leucite from a starting potash feldspar raw material occurs at temperature between 1,100 and 1,200°C and needs to held there for several hours. The glass-ceramics that is obtained generally contains less than 40 vol% tetragonal leucite within the residual glass matrix. Depending on the chemical compositions, feldspars can be heated to form the eutectic mixture of leucite and liquid according to the K_2O–Al_2O_3–SiO_2 ternary phase diagram. The amount of leucite formed on firing of potash feldspar depends on the compositional ratios of K_2O, Al_2O_3, and SiO_2 and the heat treatment temperature. Significant crystallization is thought to be related to the presence of a minimum of 1 wt% CaO and at least 12 wt% K_2O content in the glasses.

The K_2O–Al_2O_3–SiO_2 ternary phase system published by the American Ceramic Society in 1960 is shown in Fig. 10.1. The system includes various phases that can crystallize including leucite, mullite, cristobalite, tridymite, and quartz. The stoichiometric leucite is found to melt at a temperature in excess of 1,800°C and increasing the reaction temperature will increase the percentage conversion of the ingredients to leucite. Of course the use of K_2O–Al_2O_3–SiO_2 components alone is not suitable for making low fusion glass-ceramics for dental use. Other oxide components such as Na_2O, MgO, CaO, TiO_2, Eu_2O_3, Li_2O, and BaO need to be added to

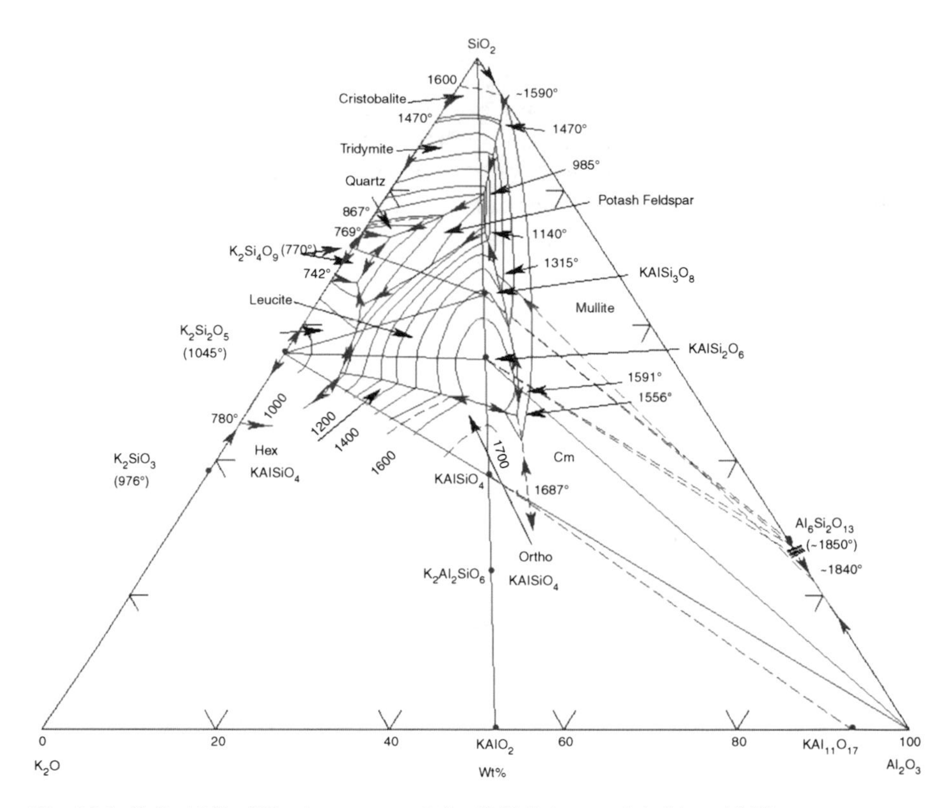

Fig. 10.1 K_2O–Al_2O_3–SiO_2 phase system (after E. F. Osborn and A. Muan 1960)

the base ingredients of alumina, silica and potassium oxide to reduce the melting temperature, at the same time as retaining the capability of creating leucite crystallinity in the glass-ceramic.

10.3 Chemical Compositions of Leucite Ceramics

The stoichiometric chemical formula of leucite phase mineral is K_2O. $Al_2O_3 \cdot 4SiO_2$. Using the stoichiometric composition to produce leucite glass-ceramics needs a very high temperature of about 1,600–1,700°C for melting, while casting or fritting of the resulted glass is difficult. So, it is very difficult to achieve a low fusion leucite ceramics without extra oxide additives either on the account of the main oxides or over the stoichiometric range. Thus, additives such as Na_2O, B_2O_3, Li_2O, BaO, P_2O_3, ZnO, etc. are generally added to the main batch compositions.

Potassium oxide enters into the silicate structure lattice as a basic component for building the leucite crystalline phase and serves to lower the fusion temperature and raise the thermal expansion coefficient. The high potassium content is essential as it tends to increase the content and stability of the leucite phase in compositions with a low glass transition temperature.

In particular, Na_2O is an important additive for obtaining a higher coefficient of thermal expansion and lower melting temperature. High alkali oxides (Na_2O and K_2O) together with Li_2O assure a relatively low glass transition temperature (Tg) and flexibility to manage the thermal expansion for improving its adaptability to coat alloys of different thermal expansion.

Li_2O, added in the order of 1–3%, is highly effective in reducing the glass transition temperature and melting temperature and enhancing the crystallinity. The incorporation of Li_2O not only functions to modify the viscosity during fusion, but also improves flowability, nucleation and grain growth, and reduces the thermal expansion coefficient.

LiF is a beneficial component to the crystallization of leucite in the starting glass composition as it acts as a powerful flux during melting to lower the viscosity and fusion temperature of the resulting glass. LiF is believed to facilitate the process of surface crystallization by the action of the Li and F ions on the glass viscosity, the surface tension and wetting of the leucite phase. In particular, a fluorine content in the order of about 0.2–1% acts to neutralize the effect of Al_2O_3, which is known to inhibit surface crystallization. The presence of Li_2O and/or LiF also assists in reducing the coefficient of thermal expansion and decreasing the fusion temperature of leucite ceramics.

CaO strengthens the glassy phase and reduces its solubility in the presence of a high K_2O content. CaO is desirable and appears to function either alone or in combination with MgO in strengthening the glassy phase. Calcium oxide can be introduced in the form of calcium carbonate and magnesium oxide can be introduced in the form of magnesite, talc or in the form of pure magnesium carbonate. A small amount of CaO of up to 2% is necessary for the good crystallization of tetragonal leucite in the glass.

Other compatible components in the glass formulation such as ZnO and Ta_2O_5 are added in the case of gold alloys to secure good bonding and develop opacity. Other oxide components can be added to adjust the final composition and properties of the leucite glass-ceramics.

10.4 The Surface Crystallization Mechanism of Leucite

When producing a leucite glass-ceramic two problems need to be addressed. First, the stoichiometric leucite based on the system K_2O–Al_2O_3–SiO_2 melts at a very high temperature, although using the stoichiometric composition is expected to yield the maximum content of crystalline leucite in the glass. Nevertheless, the stoichiometric composition is not suitable for making low fusion glass-ceramics for dental use and other oxide components need to be added to reduce the melting temperature and keep suitable leucite crystallinity in the glass-ceramic. The second problem is that

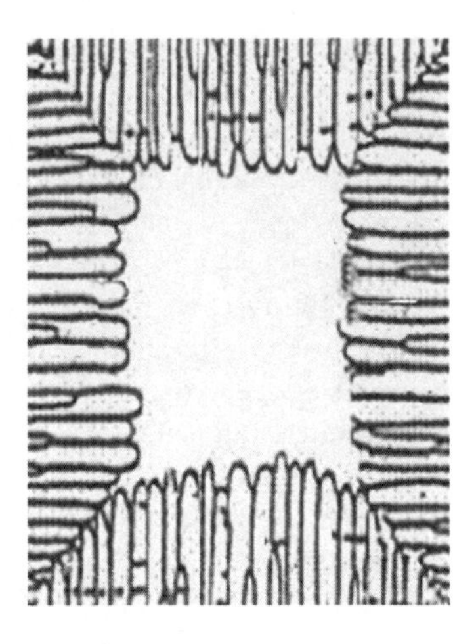

Fig. 10.2 Crystallization of crystals from the surface inward into the bulk glass

leucite glasses based on the stoichiometric composition are very difficult to crystallize from glass blocks and only unsatisfactory crystal growth may result during heat treatment of glass blocks and only after very long heat treatment schedules. The explanation is that the leucite is really only able to crystallize by a surface and not a bulk crystallization mechanism. The leucite glasses are thus best crystallized by the surface crystallization mechanism from fine glass powders, which can then be sintered to produce glass bodies.

When a monolithic bulk glass, based on leucite formulations is heat treated crystals grow from the surface of the glass inward, leaving the bulk of the glass amorphous, even in the presence of an internal nucleating agent. On increasing the time of treatment, only unusual spherulitic crystals form, resulting in a very weak crystalline structure as shown in Fig. 10.2. The surface crystallization grows significantly faster than that from the inner boundaries of the bulk glass and results in an initial formation of leucite crystals that is highly disordered.

Certain glasses simply cannot be converted into glass-ceramics by bulk crystallization and leucite glasses fall into this category. The effect of nucleation and its role in development of surface crystallization is less important than in the case of bulk crystallization, but it shares in promoting the leucite formation and growth. The surface crystallization of the glass into glass-ceramics can be done from glass powders by a combination of sintering and surface crystallization.

The surface crystallization from glass powders is primarily governed by the development and growth of leucite crystals in two dimensions starting from the surface. When a glass frit is milled into micron size powders, the total surface area of the glass becomes very high compared to that of the bulk glass. So the crystallization is achieved from individual glass particles, which become in effect separate bulk glass units. The densification is achieved by sintering the crystallized powder, where the generation of a liquid phase at temperature above the softening temperature of the glass fuses the particles together.

The crystallization reaction depends on the chemical composition and heat treatment and the increase in the surface area of the glass, each enhancing the rate of crystallization. The contact between the glass particles leads to a higher reaction rate and in turn increases the possibility of crystal development in each glass particle. The crystallization is associated with a preferred orientation of crystals that grow from the surface. In the case of leucite crystals, these develop perpendicular to the inner surface and grow to the center of the sample.

Moving far from the stoichiometric composition of leucite, by adding other oxide components to serve nucleation and crystallization, will not stop leucite from crystallizing according to the mechanism of surface crystallization or shift the mechanism toward bulk crystallization. If the classical nucleating agents TiO_2 or CeO_2 are added, the formation of the entire crystalline structure may significantly change, but it is not easy to establish the exact effect of nucleation on the primary phase formation or creating bulk crystallization. The presence of these nucleating agents may increase the number of crystals at the surface and not in the bulk of the glass without evidence of epitaxial growth.

10.5 Crystalline Structure of Leucite

Leucite can exist both as a cubic and a tetragonal crystalline structure. To understand the behavior of leucite, it is necessary to consider in detail its structure and its behavior in the course of phase transformations. Leucite ($KAlSi_2O_6$) is a potassium aluminosilicate mineral phase with a framework silicate structure.

The structure of leucite represents a continuous three-dimensional skeleton formed of (Si, Al)O_4 tetrahedra. Each of the tetrahedra shares all its oxygen ions with its neighboring tetrahedra. The tetrahedra are arranged into four-, six-, and eight-member rings. The six member rings form a sequence of parallel layers and open channels passed through in a perpendicular direction.

Within the leucite structure there exist two different cation sites. The larger site contains 16 positions coordinated by 12 oxygen ions, organized in line with channels formed by six-membered rings. The second site contains 24 positions coordinated by six oxygen ions, although only 16 are filled. The larger sites are usually occupied by potassium, rubidium or caesium ions and smaller sites by water molecules or left vacant.

Both cubic and tetragonal leucite contain 16 ($KAlSi_2O_6$) in the unit cell. Potassium cations are placed in the channels and aligned in rows parallel to the channel axes in cubic leucite. A reversible temperature dependent cubic to tetragonal leucite phase transformation occurs around 620°C. The structure of cubic leucite in Fig. 10.3 is stable above a temperature of 620°C. The elementary unit cell contains 48 tetrahedra and 16 potassium cations. At lower temperatures, the potassium ions are too small to fill the cavities in the cubic structure and moved away from their positions.

On cooling down the cubic structure is transformed into a tetragonal one with the crystal structure shown in Fig. 10.4. This conversion is rapid, reversible, and

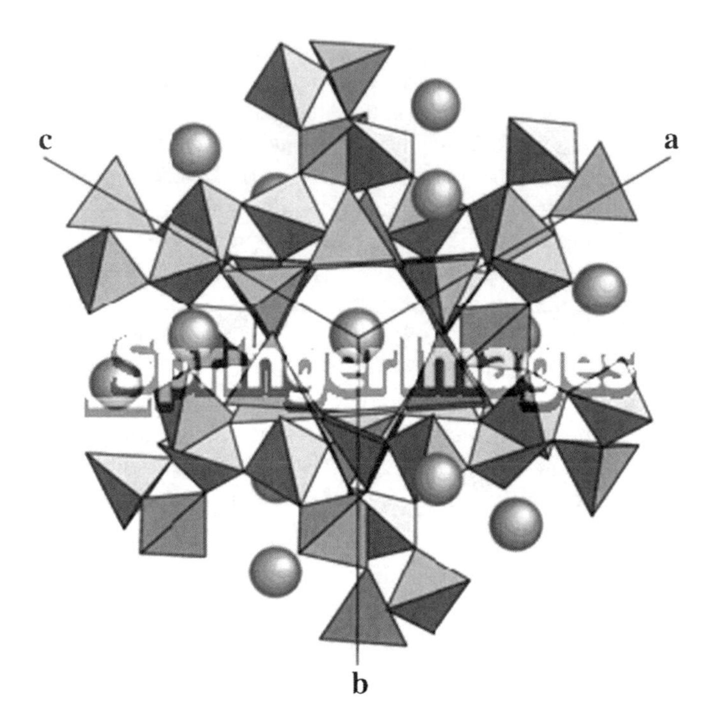

Fig. 10.3 The crystal structure of cubic leucite

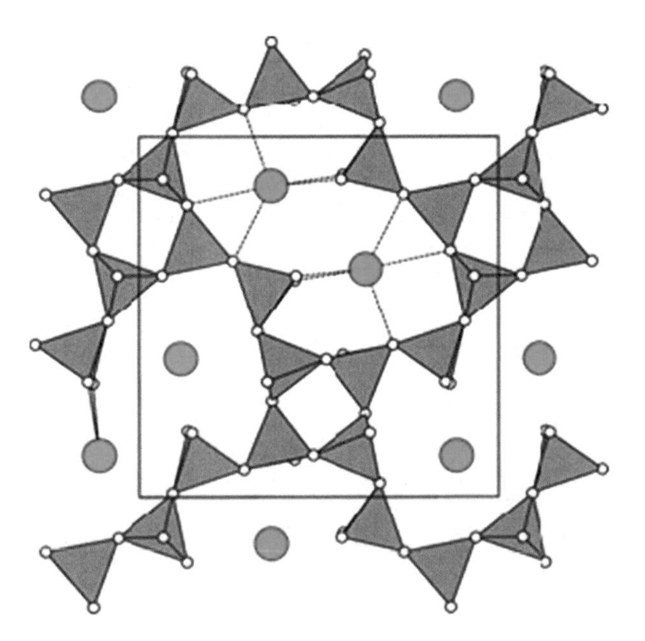

Fig. 10.4 The structure of tetragonal leucite of Mazzi et al. (1976)

continuous second-order transformation. The cubic structure of leucite has the significant property of being extremely tolerant with respect to changes in composition. K^+ cations are readily replaceable for Rb^+ or Cs^+. Aluminum cations in the tetrahedral sites can be easily substituted with boron.

By the substitution of alkali cations the stability of the cubic modification is shifted toward lower temperatures and the size of the elementary cell grows. The conversion of cubic leucite into the tetragonal modification involves deformation of the six-member tetrahedron layers. This change in the framework results in a change in symmetry and strain induced crystal twinning. Leucite transformation is associated with a reversible 1.2% discontinuous volume change, from low temperature tetragonal to the high temperature cubic leucite.

10.6 Crystalline Leucite Phases

Leucite is a valuable reinforcing phase in all ceramic restorations and enhances easy thermal compatibility for bonding to metallic substrates and ceramic cores. Leucite glass-ceramics are extremely adaptable, as they can be processed into dental prostheses via many mechanisms including, traditional sintering, hot pressing, and computer-aided design and machining. Typical leucite volume fraction for dental ceramics varies between 5 and 45% of tetragonal leucite in the glassy matrix for veneering purposes. One of the main functions of the developed leucite inclusions within the ceramics body is to act as crack stoppers.

The difference in thermal expansion between the tetragonal leucite crystals and the glassy matrix in dental ceramic is around (8×10^{-6}/°C). Compressive stresses are generated around the tetragonal crystals dispersed in the glassy matrix due to the difference in thermal expansion. On cooling, the tetragonal leucite crystals contracts more than the glassy matrix, resulting in the crystals pulling away from the glassy matrix, creating cracks around the leucite clusters.

The content and morphology of the leucite crystals in the ceramic may be modified by multiple heat treatments and cooling during fabrication. The thermal expansion mismatch between the tetragonal leucite crystals and the glass matrix developed during leucite transformation often causes signs of microcracking around leucite crystals. The size of the microcracks can be reduced by reducing the leucite crystal size, resulting in a toughening effect rather than microstructural failure. The nucleation and crystallization of aluminosilicate glasses in the K_2O–Al_2O_3–SiO_2 system is very important in reducing the crystal size, control the leucite morphology, distribution and the volume fraction.

The transformation of tetragonal leucite to a *cubic leucite* at 625°C is anisotropic (variable in degree or level in different directions) and proceeds diffusion-free, by the slipping or twinning mechanism. In the case of leucite, the martensitic transformation is facilitated by sliding of the six-member tetrahedron ring planes. The incorporation of leucite in the glassy matrix represents a promising way of reinforcing this glass-ceramic. What is still required is a better understanding of the

Table 10.1 Glass chemical compositions that cystallize into leucite glass-ceramic

Oxides	X0	X1	X2
K_2O	11.00	13.32	13.80
Na_2O	4.20	4.28	3.42
Al_2O_3	13.60	14.48	15.00
SiO_2	61.00	56.21	55.44
CaO	1.70	1.28	1.02
MgF_2	0.00	0.74	1.18
B_2O_3	2.50	3.88	4.70
MgO	0.00	0.96	1.53
TiO_2	6.00	4.88	3.90

mechanisms of promoting fracture toughness, about which very little is known so far in contrast to the martensitic transformation in zirconia ceramics.

10.6.1 Tetragonal Leucite Glass-Ceramics

To make a reliable veneering ceramic for coating metal substrates with a thin glass-ceramic film, many of the technical problems can be solved by a reduction in the difference in the thermal expansion coefficients of the ceramic coating and the metal substrate. The crystallization of tetragonal leucite in the system K_2O–Al_2O_3–SiO_2 during the cooling raises the thermal expansion coefficient of the ceramics and when carefully controlled, to a value close to the expansion of metal substrates. Thus, according to the type of the metal substrate, the thermal expansion of the glass-ceramic coating can to be adjusted to a certain value.

Tetragonal leucite is a critical component in glass-ceramics to obtain the correct thermal expansion matching. In addition tetragonal leucite, can impart higher strength, greater durability and the desired translucency to the final ceramic. A considerable way to manage the required properties of glass-ceramics can be made by mixing different frits, processed under different conditions and with different chemical compositions, to obtain the overall desired chemical, mechanical and aesthetic characteristics.

Three glass compositions that crystallize into glass-ceramics containing tetragonal leucite are shown in Table 10.1. The XRD of tetragonal leucite glass-ceramic is shown in Fig. 10.5. The dispersion of crystals of leucite will exhibit higher fracture toughness than the same glasses free from leucite crystals.

The addition of lithium oxide can promote crystallization, while the addition of sodium significantly suppresses the tetragonal leucite crystallization. These observations opened the way to the development of dental ceramics containing various levels of leucite content, which made it possible to prepare leucite ceramics that are readily applicable for coating various high expansion metal alloys. The range of the desirable properties of tetragonal leucite glass-ceramic required for coating metal alloy is presented in Table 10.2.

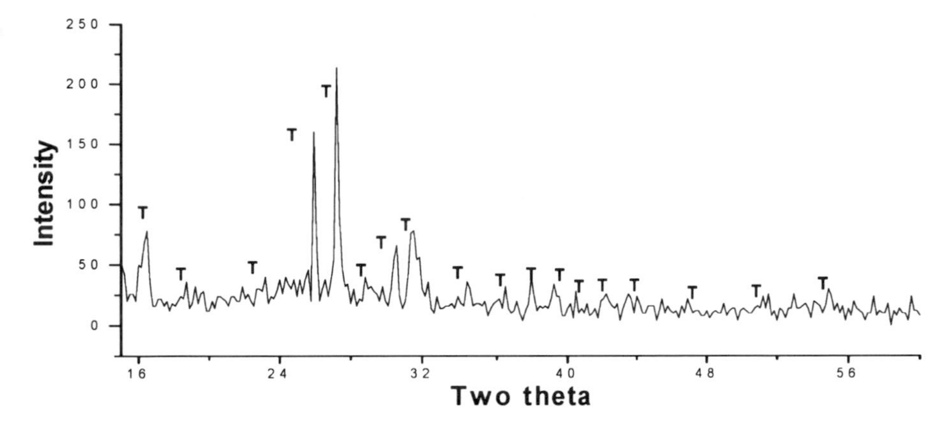

Fig. 10.5 XRD analysis of leucite body fast fired at 950°C for 2 min

Table 10.2 Properties of tetragonal leucite glass-ceramics

Property	Unit	Value
TEC (25–500)	10^{-7}/°C	135–145
Softening point	°C	640
Solubility	μg/cm^3	12
Density	g/cm^3	2.4
Bending strength	Mpa	90
Bond strength	Mpa	50
Modulus of elasticity	Gpa	95
Vickers hardness	HV	500
Natural enamel hardness	HV	400–500
Grain size of leucite	μm	1–2

10.6.2 Cubic Leucite Glass-Ceramics

The high coefficient of thermal expansion of tetragonal leucite reinforced ceramics makes it suitable for veneering metal alloys. But mechanical strength and the fracture toughness are not good enough to be used as core ceramics. Any trial to improve its mechanical strength through the increase in the content of leucite in the glass-ceramics will result in further increase in the thermal expansion without significant improvement, in addition to losing the low fusion quality. So, the tetragonal leucite glass-ceramics has a limited strength to be used in the construction of all ceramics dental cores as an alternative to metal substrates. The cubic leucite, which is also called the high leucite phase, may be used in the construction of dental core for all dental ceramics.

The crystallization of cubic leucite is found to improve the strength of the ceramics to nearly double the strength generated by the tetragonal leucite. In addition, the cubic leucite content results in relatively lower thermal expansion glass-ceramics compared to that reached by the tetragonal containing glass-ceramics. The formation

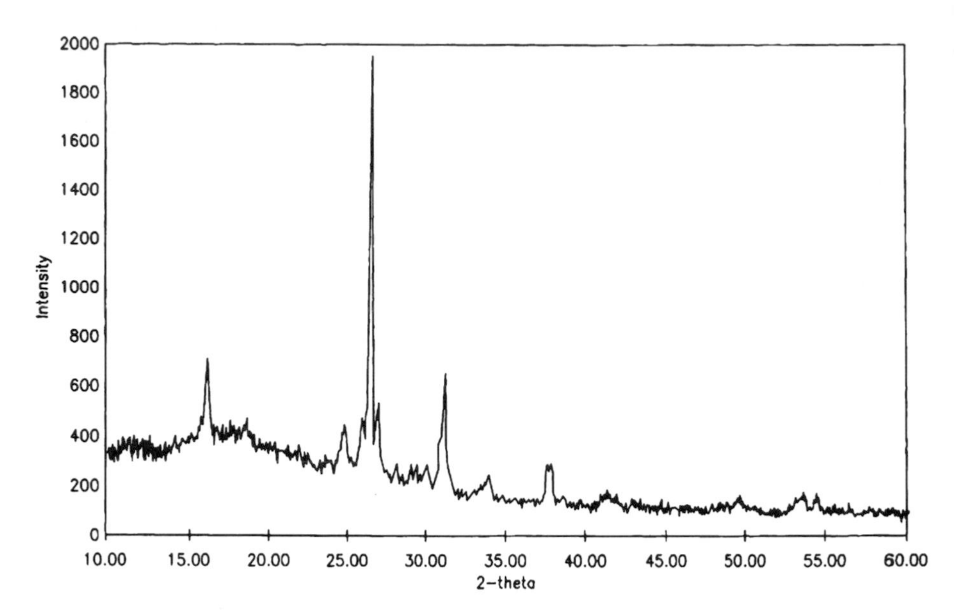

Fig. 10.6 XRD of cubic leucite

of cubic leucite relies on the presence of more than 18% K_2O in the composition and on the type of crystallization mechanism. The crystallization of cubic leucite in the high viscosity $K_2O–Al_2O_3–SiO_2$ system can be assisted by using titania, zirconia or P_2O_5 as catalysts. The cubic leucite phase may be produced as the only crystalline phase on long heat treatment times and high temperature from the powdered glass. No cubic leucite can be formed from the bulk glass, showing that cubic leucite is produced by the surface crystallization mechanism.

A small amount of CaO, approximately 1 wt%, is needed to stabilize the cubic phase at room temperature, when the crystallization is achieved from the high viscosity leucite $K_2O–Al_2O_3–SiO_2$ system.

Another significant way to crystallize cubic leucite, is to incorporate about 2-mol% of Cs_2O in the glass composition close to that of the high-strength feldspathic ceramics. The role of caesium is to stabilize the cubic leucite phase at ambient temperatures. The cubic leucite phase is formed during the crystallization following melting and is retained in the glass-ceramic at room temperature. The high temperature frit formation can produce varying amounts of the cubic phase of leucite as a metastable phase. The XRD pattern of cubic leucite is shown in Fig. 10.6.

The cubic glass-ceramic is too refractory to be processed in the dental laboratory as the fusion temperature exceeds 1,200°C, which is considered a high fusion temperature. Also, the grain size and the microstructural uniformity and the distribution of crystals are not adequate to produce sufficiently strong ceramics.

10.6.3 Pollucite Glass-Ceramics

Another process to adjust the thermal expansion coefficient is to introduce a pollucite phase into the leucite ceramic. Pollucite is a relatively low thermal expansion cesium-aluminosilicate glass-ceramic, which has the same cubic structure as high-leucite at room temperature. The addition of less than 2% cesium oxide will result in the stabilization of the low expansion cubic leucite phase at room temperature. However, if the crystallization of pollucite is desired then one needs a higher content of Cs_2O to forms a continuous series of solid solutions with leucite. As the cesium level in the leucite ceramic is increased, the thermal expansion coefficient decreases. The leucite/pollucite assumes the high leucite cubic structure at room temperature, and the coefficient of thermal expansion continues to decrease with increased Cs_2O content.

Aside from the presence of Cs_2O in glass compositions, the chemical composition of the starting glass should be reasonably close to the composition of high-strength feldspathic glass-ceramics. The cubic leucite phase is formed during the crystallization heat treatment following melting and retained in the glass-ceramic at room temperature. Although it is a simple and cost-effective method, it does not provide sufficient control of the microstructure and morphology of the developed leucite glass-ceramic.

10.7 Criteria for Choosing the Compositions of Ceramic Coatings

The basis for the development of low fusing glass-ceramics is dependent on a strict control over the chemical composition and the various processing conditions. The combination of the metal alloy with the leucite glass-ceramics imposes special requirements on both materials. The requirements the ceramics must satisfy are as follows:

1. Good flowability during fusion.
2. The melting interval is sufficiently lower than the firing temperature of the metal alloy.
3. Linear thermal expansion coefficient is preferably slightly lower than that of the metal alloy.
4. Adequate ceramic-metal bond strength.
5. Adequate strength of the ceramics to resist the very high forces in the occlusal area.
6. Adequate chemical resistance to suit the use in the oral cavity.
7. Adequate microhardness to match that of the natural tooth.
8. Avoid the destructive transformations during cooling specially those associated with higher volume change as the transformation of cubic to tetragonal leucite.
9. The ability of the ceramics to be coated with very thin layer without losing its character.

10.8 Design of Glass-Ceramic Veneers for Metal Substructures

A widely used technique for making a dental prosthesis is the ceramic fused to metal restoration. The structure of a synthetic dental crown is shown in Fig. 10.7. The method is popular and provides a strong enough and durable crown or bridge and the marginal fit of the restorations is considered excellent. The method also provides a tight impervious ceramic coating having the appearance of natural dentition. The metal substructure provides enhanced structural strength, durability, and toughness to the restoration.

The process of veneering the metal involves applying multiple layers of ceramic with similar thermal expansion coefficients and different optical properties including an opaque ceramic layer, dentin ceramics layer and enamel ceramics layer. The sequence of laying down a veneer on a metal substructure is shown in Fig. 10.8. The opaque ceramic is first applied over the metal framework to form an opaque layer that masks the metal color. An aqueous slurry of the opaque ceramic is prepared and applied to the metal substrate and then fired at temperature between 880 and 950°C for 1 min. It is important that the maturing temperature of the opaque ceramics layer is at least 100°C below the solidus temperature of the metallic framework.

It is necessary that the opaque ceramics exhibits a coefficient of thermal expansion (CTE) substantially equal to or slightly above the CTE of the metal alloy to which the ceramics is being applied. The coefficient of thermal expansion must be equal to or slightly greater than that of the veneering ceramics so that no destructive cracks are produced in the coating layer during firing and cooling cycles. The thermal compatibility and thermal stability of the opaque coating can be adjusted via

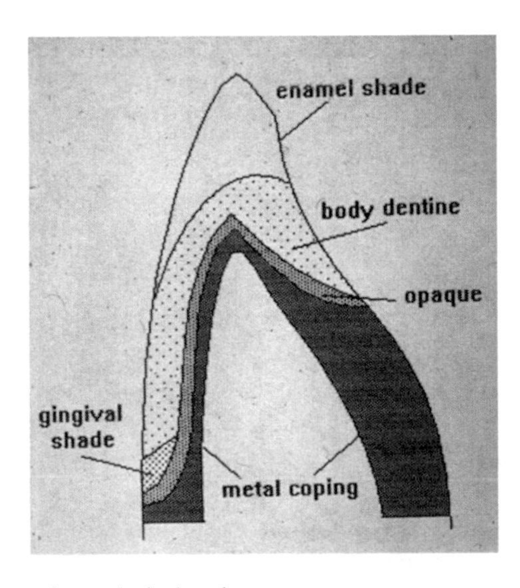

Fig. 10.7 The structure of a synthetic dental crown

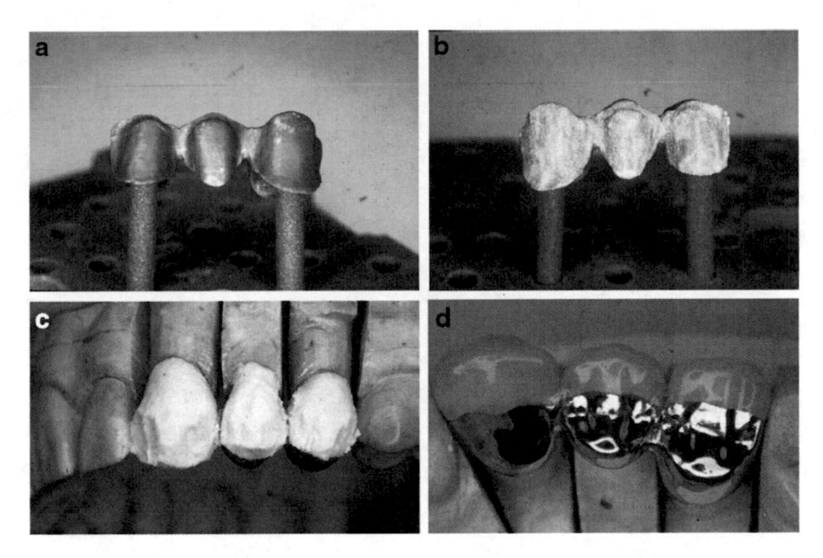

Fig. 10.8 Sequence of laying down a veneer on a metal substructure: (**a**) metal casting, (**b**) opaque layer, (**c**) build up with dentin/enamel shades, and (**d**) final restorations (supplied by Dr. T Johnson, School of Clinical dentistry, University of Sheffield)

modification of the chemical composition to ensure that the ceramic is bonded to the metal substructure with no visible cracks after multiple firings of the subsequently applied ceramic veneering layers. If the opaque layer delaminates from the metal substructure, the opaque coating is not considered thermally stable with the overlying ceramics layer.

A dentin ceramics layer formed from a mixture of base ceramics, stained ceramics, and fluorescing agent is subsequently applied to build up the restoration. The restoration will be fired again between 850 and 910°C to mature the dentin layer, which must be thermally compatible with the opaque layer underneath and the subsequently applied overlying enamel coating, with no cracks visible in the dentin layer. An enamel layer is finally applied over the dentin body layer and the layer is fired at a temperature in the range of 800–850°C. The ceramic coating forms a hard and durable layer and the final restoration has a shade that matches the shade and translucency of the patient's natural teeth. The resulting layer of the ceramic must be thermally compatible and thermally stable with no substantial visible cracks.

10.9 How to Modify Thermal Expansion Coefficient of Ceramic Coating

The thermal compatibility is the most important requirements for a reliable veneering glass-ceramic fused to metal restorations. Thermal compatibility of ceramic fused to metal over the range from room temperature to the glass transition temperature

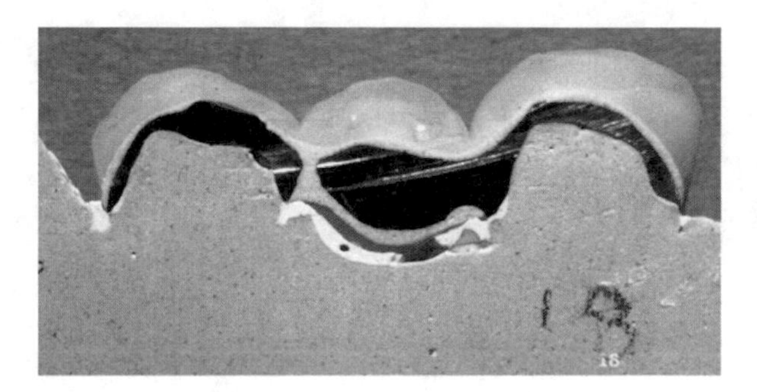

Fig. 10.9 Sectional view of a ceramic veneer bonded to a metal framework for a three-unit dental bridge

can be assessed by measuring the expansion coefficients of metal and glass-ceramic in the range of ~20–600°C.

When a dental ceramic is to be used as a veneer on a metal substructure, then the thermal expansion must be controlled to ensure good bonding of the ceramic to the metal of the dental prosthesis. A sectional view of a ceramic veneer bonded to a metal framework for a three-unit dental bridge is shown in (Fig. 10.9).

A slight difference in thermal expansion coefficient is an important parameter in metal-ceramic compatibility. The alloy must have a slightly higher thermal expansion coefficient than the ceramic to create compressive stress in the ceramic during cooling. The adaptation of the thermal expansion coefficient of glass-ceramic to the desired level, which is close to and slightly lower than the metal alloys, can be monitored by the crystallization of leucite and its crystalline content in the ceramic body.

The crystallization of tetragonal leucite imparts an increase in the thermal expansion through the dispersion of the high expansion tetragonal leucite in a glassy matrix. Different methods can be used to modify the thermal expansion coefficient to match the metal substructures.

One technique of thermal expansion modification is to prepare a glass composition holding various characteristics. Using this technique, it is possible to modify the expansion coefficient through modifying the chemical composition, and the crystalline content in the glass matrix. The content of the crystalline leucite determines the thermal expansion coefficient and maturing temperature and it is possible to achieve ceramics with a wide range of thermal expansion between 50 and $200 \times 10^{-7}/°C$. First, the glass composition should be able to fuse at a temperature lower than that of the melting temperature of the metal substructure by more than 100°C. Second, the glass should crystallize to produce tetragonal leucite crystals that are able retain their crystalline structure while being fused onto the metal substrate during the processing of the restoration. Third, the developed glass-ceramic should be able to mature in few minutes (1–3 min) and considered a low fusion glass-ceramic. Fourth, the developed crystalline tetragonal leucite crystals should have a thermal expansion coefficient near or similar to that of the hosting glass matrix to avoid excessive cracking around the leucite crystals. Fifth, the overall thermal expansion of the developed leucite glass-ceramic should be slightly lower than the thermal expansion coefficient of the supporting metal alloy.

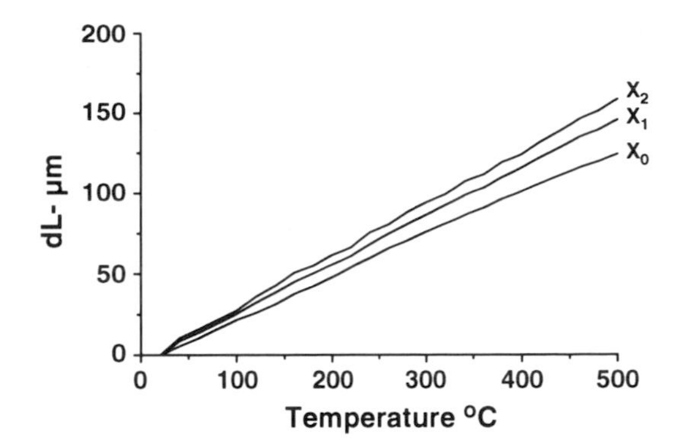

Fig. 10.10 Thermal expansion modification

This method of TEC adjustment gives a reliable adjustment and stability with a very small variation in the thermal expansion coefficient over repeated firings. Figure 10.10 shows the thermal expansion modification of three glasses with chemical composition in Table 10.1 that crystallize into tetragonal leucite glass-ceramic.

A second technique of thermal expansion modification is to adjust the thermal expansion coefficient by means of a combination of two or more different frits with variable thermal expansion coefficients, which have been previously prepared. The frits should be low fusion glass-ceramics with high thermal expansion stability. The technique depends on using two main frits. One of these frits is known as a high fusing temperature leucite frit in which leucite crystallizes out of a glass by heat treatment leaving a matrix, which is relatively deficient in aluminum oxide and potassium oxide.

The high leucite frit will be responsible for the development of tetragonal leucite crystals in the veneering glass-ceramics. The second frit is made of a low fusing leucite glass composition with a composition closely similar to the glass matrix of the high leucite glass frit and its main objective is to reduce the fusion temperature.

It is important that there is an absence of any interaction between the two frits to achieve the required thermal expansion coefficient. The idea is fine to the extent that the thermal expansion can be adjusted easily, but the mechanical properties could be affected. Through mixing both a high melting glass frit and a low melting frit, it is not easy to reduce the thermal expansion coefficient without lowering the firing temperature of the final ceramic and controlling the maturing temperature is difficult. Adding an excess of low temperature glass frit with respect to the high leucite frit, results in an unacceptable product because the low melting glass frit acts very aggressively on the high leucite frit, the result being an unstable product. The glass frit needs to be slightly different from the matrix of the high leucite frit and the addition of oxides is needed to adjust the properties rather than the thermal expansion coefficient.

A third glass frit complementary to the high temperature leucite frit and a low melting frit can also be used to adjust the thermal expansion coefficient by holding the melting point without a significant change. The low fusion frit should have a composition that considerably decreases the melting point and yet does not reduce the thermal expansion too much.

Table 10.3 EDX analysis calculated in oxide wt% made for the microstructural features in Fig. 10.11

Oxides	A	B	C
Na_2O	1.71	1.35	4.32
MgO		0.80	0.69
Al_2O_3	18.53	12.31	15.33
SiO_2	61.46	69.05	64.26
K_2O	16.34	7.80	10.53
CaO		2.60	0.89
BaO	1.97	3.68	2.12
TiO_2		0.41	1.81

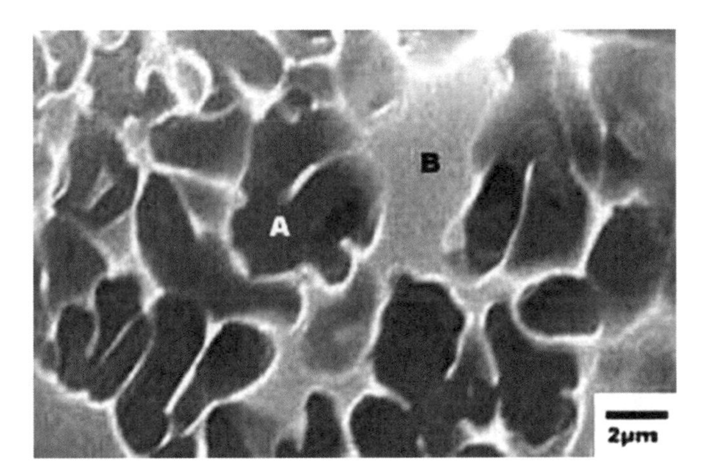

Fig. 10.11 Tetragonal leucite colonies showing the position of the point EDX analyses

Although the above combination of frits has been used with some success in thermal expansion adjustment, it often results in ceramics variation in terms of other properties during repeated firing. However, it is very important to provide ceramics, which shows a high stability in expansion. Sometimes, the use of more than one frit in adjusting the TEC may impair the other properties such as the microhardness or the chemical resistance. So, the scientist must be careful to use adequate mixtures and chemical components to produce reliable products when using these different approaches.

A composition satisfying good conditions for thermal expansion adjustment is found in a frit having a relatively high sodium oxide and high potassium oxide content together with a relatively low aluminum oxide content. The low fusion frit as compared with the conventional complementary glass frits has a melting point lower than about 700°C and the thermal expansion coefficient is typically about 130×10^{-7}/°C. The low temperature frit aims to reduce the firing temperature and adjust the thermal expansion coefficient. Table 10.3 shows a chemical analysis based on EDX data, for an optimized tetragonal leucite depicted in Fig. 10.11. The positions of the EDX analyses are presented as (A) leucite tetragonal crystals, (B) leucite glass matrix, and (C) the whole picture area.

10.10 Opacity Development in Veneering Glass-Ceramics

Opaque porcelains are those sufficiently low in light transparency as to hide the substrate from view. They are usually white but this is not necessarily a requirement. The more opaque is the ceramic, the greater the hiding power and the thinner the thickness that needs to be applied. The opacity is a consequence of the refraction and reflection of light by phases and particles suspended in the clear glassy matrix. The opacity can be achieved by as small a difference as 0.05 in refractive index of the particles from the glass.

The degree of opacity depends on several factors:

1. The difference in refractive indices between the transparent matrix and the opacity causing dispersed particles.
2. The particle size and the degree of dispersion of the dispersed phase; the smaller the particle size and more uniform the dispersion, the greater the opacity.
3. The opacity increases as the number of reflecting particles per unit volume increases.
4. The character and the nature of grain boundaries between the opacifying particles and the hosting matrix. The sharper the boundaries, the more effective the opacifiers.
5. The solubility of the opacifier in the matrix is a crucial factor; the greater is the solubility of the opacifier, the more the corrosive the matrix, and the lower the hiding power and opacity.

The route for controlling the opacity in a ceramic is by the following:

1. The used of fine particle size and high refractive index opacifiers.
2. Utilization of liquid–liquid phase separation through composition modification and heat treatment such as those achieved in mica glass-ceramics by the use of fluorides.
3. Precipitation of a crystalline phase or phases with higher or lower refractive index than the glass.

The intensity of opacification is related to the difference in the index of refraction of the glass from that of the opacifier.

$$\text{Intensity of opacification} = \frac{I_e(n_2 + n_1)^2}{(n_2 - n_1)^2},$$

where
I_e = incident light,
n_1 = index of refraction of the glass matrix,
n_2 = index of refraction of the dispersed phase.

Table 10.4 Refractive indices and melting temperatures for a range of opacifying oxides

Oxide	Refractive index	Melting point (°C)
ZrO_2	2.20	2,715
TiO_2	2.60	1,830
SnO_2	2.10	1,625
Sb_2O_3	2.30	655
P_2O_5	1.60	570
MoO_3		795
ZnO	2.10	1,975

Opaque glass-ceramics can be manufactured in two ways. The first way is to add opacifiers to the frit during milling as mill additions. The most common dental opacifiers are based on their insolubility in the glass. These opacifiers are listed in Table 10.4.

The second type is to develop crystalline phases with refractive indices different from the surrounding glassy matrix. The opacity increases with the increase in the dispersed crystalline content. For example, the development of spodumene or mica crystalline phases increases the opacity and reduces the translucency depending on the content of the crystalline phase in the glass.

In fluorine containing glass-ceramic systems, fluorides such as aluminum fluoride, barium fluoride, magnesium fluoride, and calcium fluoride, under proper circumstances, serve as effective opacifiers if the conditions of heat treatment and the chemical composition are adjusted.

Tin oxide is the traditional opacifier for glasses. Tin oxide is still not favoured and other oxides such as zirconia and titania give better opacity and brilliance. Alternatively, P_2O_3 is comparable to tin oxide in terms of opacification with respect to the color, texture, and brilliance of the product, but works in a different temperature range and different glass chemical compositions in which they may act as opacifiers. Silica in the glass may also serve as an opacifier, but it is very difficult to control or use in the case of dental materials.

Titania is also a powerful opacifier and is used in many glass applications. Titanium oxide may act as a white opacifier only at low temperatures, but in the case of low fusion dental ceramics, it will develop a yellowish white color that disturbs the gloss and brilliance of the developed dental crown. The grain size from submicron to 20 μm is appropriate for a TiO_2 opacifier.

ZnO is a desirable opacifier in dental ceramics as it gives a very nice gloss and controlled opacity. This is in addition to its role as a factor for increasing the chemical resistance and fluxing power. ZnO can be used in amounts of less than 5% and also can be utilized for blocking the formation of quartz as it aids in the dissolution of the silica during fusion.

10.11 Microstructural Optimization of Low Fusion Leucite Ceramics

The low fusion tetragonal leucite glass-ceramic is considered the backbone for most dental veneering ceramics. For this leucite glass-ceramic to be used in the manufacturing of veneers, it should satisfy certain criteria namely thermal expansion coefficient that matches the utilized substrate, low fusion temperature that permit the maturing of the veneers in few minutes to suit the production flow, good mechanical strength to withstand the occlusal pressure, microhardness like that of the natural tooth, good bond strength with the substrate and refractive index similar to that of the parent glass. To make a good compromise among all of these properties without hindering any of the properties is a very complex process. Obtaining a uniform microstructure with a homogenous distribution of the leucite crystals in the glass matrix without cracking or any structural inhomogeneity is the most crucial key in achieving this goal.

The leucite glass-ceramic crystallizes at a relatively low temperature of 800–950°C. The leucite crystallizes in different forms within well distributed colonies, dispersed in the glassy matrix. The crystallization starts first at the boundaries of the colonies by a surface crystallization mechanism within minutes of the soaking time.

The tetragonal leucite exists in colonies of crystals, made of acicular crystals and rod like crystals as shown in Fig. 10.12. The acicular grains vary in size with the change in composition. Acicular grains are dispersed randomly in the glassy matrix acting as a reinforcement factor. With further magnification, the fine granular grains, found to be in the acicular or rod grains, appear to reinforce the glass-ceramics as

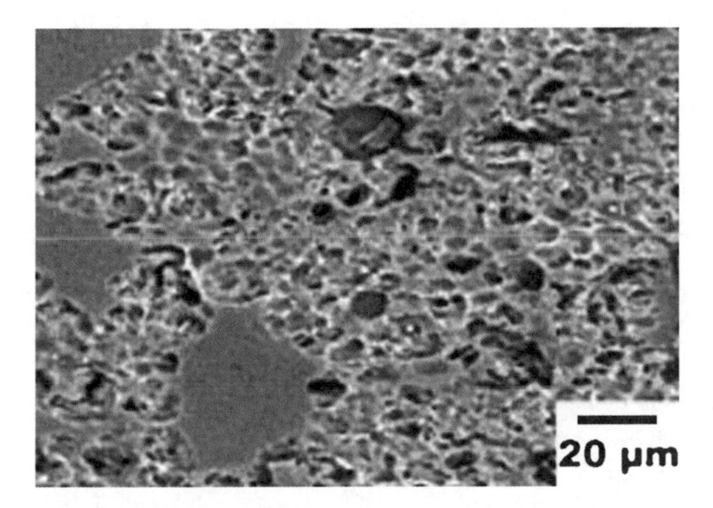

Fig. 10.12 Colonies of tetragonal leucite-like honeycombs

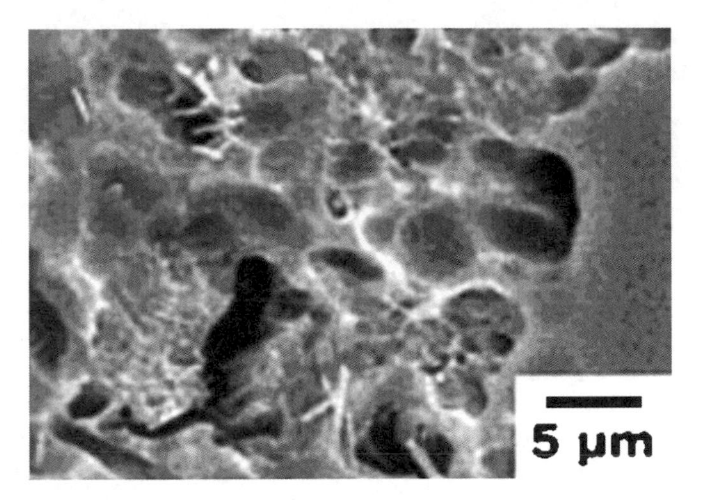

Fig. 10.13 Higher magnification of leucite colonies showing acicular tetragonal leucite crystals

shown in Fig. 10.13. The mechanical properties are improved with a machinability closely matching that of the natural teeth. No parasitic phases are formed. A uniform microstructure with fine tetragonal leucite well dispersed in the glassy matrix is apparent. The content of leucite increases with the presence of one or more of the optional components such as MgO, MgF_2 and TiO_2.

10.12 Classification of Leucite Dental Glass-Ceramics

According to the classification of dental ceramics, the low fusing ceramics are able to achieve complete densification at firing temperatures between 800 and 1,050°C in minutes. Low fusing leucite dental ceramics densify from a starting glass powder at a temperature lower than 950°C in a matter of several minutes of sintering through a fast firing treatment using a heating rate reaching more than 100°C/min.

Another classification of the leucite ceramics is more relevant to a technical and industrial approach. The classification includes three groups of leucite ceramic materials:

- Glass-ceramic for veneers for metal substrates.
- Glass-ceramic for veneers for ceramic substrates.
- Resin-bonded glass-ceramic restorations.

The veneering glass-ceramics are manufactured based on the thermal expansion needed to match either metal alloy substrates or ceramic substructures. No such constraint is imposed on the other applications of leucite ceramics.

10.13 Glass-Ceramic Veneers for Metal or Ceramic Substructures

Leucite based veneering ceramics include:

- Veneering ceramics perfectly matched with high gold content and non-precious metal alloys with a coefficient of thermal expansion (CTE) in the range of $13–15 \times 10^{-6}$/K (25–500°C).
- Veneering ceramics for coating high expansion alloys in the CTE range of $16.0–17.5 \times 10^{-6}$/K (25–500°C). These leucite ceramics are perfectly matched with gold–palladium–silver alloys. The structure of this ceramic exhibits a particularly homogeneous distribution of the crystals in the glass phases.
- Veneering ceramics for coating alumina and spinel ceramic substrates with a CTE in the range of $7.2–7.9 \times 10^{-6}$/K (25–500°C).
- Veneering ceramics for coating titanium and yttria stabilized zirconia substructure made of fine-structure feldspar ceramic with a CTE of approx. $9–10.5 \times 10^{-6}$/K (25–500°C).

10.13.1 Low Fusion Leucite Glass-Ceramics for Coating Gold Alloys

Amongst metal alloys, high gold content alloys are unique because of their appearance, especially when the ceramic does not fully cover the metal framework. They also have an excellent clinical track record. The glass-ceramics for veneering gold alloy substructures must have a relatively high thermal expansion coefficient, between 14.5 and 16.5×10^{-6}/°C (20–500°C) and a relatively low maturing temperature below 900°C.

It is quite a challenge to meet all the requirements of thermal expansion, low maturing temperature and thermal stability and compatibility. An example of three different frits that may be combined to produce a desirable veneering glass-ceramics for coating gold alloys are shown in Table 10.5.

Table 10.5 Three frits can be used in wt% proportions to produce one veneering glass-ceramic compatible with a gold alloy

Oxides	High fusion	Glass frit	Low fusion
SiO_2	63	69	70
Al_2O_3	18	12	3
K_2O	16	6	10
Na_2O	3	8	10
BaO	0	3	4
Li_2O	0	2	3
Heat treatment	900°C/1 h		
Thermal exp. ×10–6/°C (20–500°C)	18.5	9.8	13
Proportion wt%	70	20	10

10.14 Yellow Coloration in the Leucite Ceramics

Dental ceramics based on leucite glass-ceramics are found to suffer from the phenomenon of being changed to yellowish or even a greenish yellow color. This happens when the ceramics is fired to bond with the metallic substrates containing particular metal components, particularly silver metal (Ag). The coloration can also results from the use titanium oxide (TiO_2) as an opacifier. This phenomenon is referred to as yellow coloration, which is critical to the dental ceramics aesthetics and needs to be conditioned carefully during processing.

Titania has four major oxides on the base of their oxygen content: TiO, Ti_2O_3, TiO_2 and TiO_3. It is very easy for TiO_2 to lose some of its oxygen content to become $TiO_{1.995}$. The color darkens when the oxygen is less than the content of oxygen in the stoichiometric TiO_2. Titanium oxide may act as a white opacifier only at low temperatures, but in the case of low fusion dental ceramics, it will develop a yellowish white color that disturbs the gloss and brilliance of the developed glass-ceramic restoration. The yellowish white color is often due to the grain growth of the titania in conditions of excessive fluidity that encourage the formation of coarse grains of titania. This problem can be avoided by balancing the content of fluxes and silica and alumina.

The yellowish color is known to be excessive with the use of titania in rutile form, while in the anatase form it is possible to use it as opacifier with faint yellowish color. So there has been a tendency toward the use of anatase as opacifiers. Also the gloss of low fusion dental ceramics in which titania is used as opacifier is not ideal.

The finer the titania grain size the better the opacity and a grain size from submicron to 20 μm is considered to be most appropriate. It may be necessary to add up to 20% of the opacifier to achieve a thin layer of ceramics with a very high hiding power. Nevertheless, the difficulties with titania makes titania an unsuitable opacifier in dental ceramics in spite of its commercial success in other areas.

The yellow coloration made by silver is mostly due to the volatilization of the silver from the metallic frame or from the interior of the furnace contaminated with silver. Yellow coloration by silver can be diminished to some extent by a firing the dental ceramics alongside a carbon plate in the furnace. It is not possible to prevent yellow coloration completely with firing in the presence of carbon. The yellow coloration changes to creamy due to the reducing action caused by CO gas. In addition, air bubbles tend to be produced in the ceramics in that atmosphere, which in turn will reduce the bond strength. If a metal framework is produced using alloys free from silver, the ceramics fired to bond to the metal frame is no longer subject to the yellow discoloration. Using nitrate compounds as raw materials in the glass batch may also result in a yellow coloration on firing for the first time.

It continues to be a challenge to provide dental ceramics without yellow coloration under repeated firing. The best solution is to avoid the use of silver in the alloy or the use of titanium oxide or nitrate compounds in the batch. Another solution is to use fining agent such as 0.1–0.3 wt% of antimony trioxide in the ceramics composition that may result in dental ceramics free of yellow coloration upon firing. Alternatively, a fine particle size Sb_2O_3 can be added to the ceramics material, which acts as a sort of oxidizing agent and keeps the silver in an Ag-ion state to avoid yellow coloration.

Further Reading

Brodkin, D., Panzera, C.: Cubic leucite-containing dental porcelains. US-Patent 6,090,194, 2000

Brodkin, D.G., Panzera, C.: Machinable ceramics compositions and mill blanks thereof. US-Patent 035215 (A1), 2010

Brodkin, D., Panzera, C., Panzera, P.: Cubic leucite containing dental ceramics. WO-Patent 1891099, 1999

Brodkin, D., Panzera, C., Panzera, P., Pruden, J., Kaiser, M., Brightly, R.: Dental restorations. US Patent 6,155,830, 2000

Brodkin, D., Panzera, C., Panzera, P.: Machinable leucite-containing porcelain compositions and methods of manufacture. US Patent 6,133,174, 2000

Brodkin, D., Panzera, C., Panzera, P.: Dental porcelains. US-Patent 6,428,614, 2002

Brodkin, D., Gamarnik, M.: Dental restorations using nanocrystalline materials and methods of manufacture. US-Patent 7,655,586, 2010

Burk, B., Burnett, A.: Leucite-containing porcelains and method of making same. US-Patent 4,101,330, 1978

Burk, B., Burnett A.P.: Leucite-containing porcelains and method of making same. US-Patent 4,101,330, 1978

Burk, B., Windsor, C.T., Burnett, A.P., Milford, D.E.: Leucite-containing porcelains and method of making same. US-Patent 4,101,330, 1978

Christopher, C. Banasiak, S.: Ceramics system for a dental prosthesis. WO-Patent 148558 (A2), 2009

Christopher, C. Banasiak, S.: Translucent veneering for a dental prosthesis formed by a press to metal process. US-Patent 274995 (A1), 2009

Cummings, K.M., Rolf, J.C., Rosenflanz, A., Rusin, R., Swanson J.: Use of ceramics in dental and orthodontic applications. US-Patent 7,022,173, 2006

Denry, I.L.: Low expansion feldspathic ceramics. US-Patent 5994246, 1997

Denry, I.L., Mackert, J.R., Holloway, J.A., Rosenstiel, S.F.: Effect of cubic leucite stabilization on the flexural strength of Feldspathic dental ceramics. J. Dent. Res. **75**(12), 1928–1935 (1996)

Denry, I.L., Holloway, J.A., Rosenstiel, S.F.: Crystallization kinetics of a low-expansion feldspar glass for dental applications. J. Biomed. Mater. Res. **41**(3), 398–404 (1998)

Denry, L., Holloway, J., Colijn, H.: Phase transformations in a leucite-reinforced pressable dental ceramic. J. Biomed. Mater. Res. **54**(3), 351–359 (2001)

El-Meliegy, E.M.: Preparation and characterization of low fusion leucite dental ceramics. Br. Ceram. Trans. **102**(6), 261–265 (2003)

El-Meliegy, E.M.: Low fusion fluorophlogopite-leucite containing porcelain. Br. Ceram. Trans. **103**(5), 231–234 (2004)

Frank, M., Schweiger, M., Rheinberger, V., Hoeland, W.: Leucite-containing phosphosilicate glass-ceramic. US-Patent 5,698,019, 1997

Garcia-Guinea, J., Correcher, V., Rodriguez-Badiola, E.: Analysis of luminescence spectra of leucite ($KAlSiO_4$). Analyst **126**(6), 911–916 (2001)

Hermansson, L., Carlsson, R.: On the crystallization of the glassy phase in whitewares. Trans. J. Br. Ceram. Soc. **77**, 32–35 (1978)

Hermansson, L., Kraft, L., Engqvist, H., Hermansson, I., Ahnfelt, N., Gomez-Ortega, G.: Powdered material and ceramic material manufactured therefrom. US-Patent 7,351,281, 2008

Holand, W., Frank, M., Rheinberger, V.: Surface crystallization of leucite in glasses. J. Non Cryst. Solids **180**, 292–307 (1995)
Holand, W., Rheinberger, V., Apel, E., Hoen, C., Holand, M., Dommann, A., Obrecht, M., Graf-Hausner, U.: Clinical applications of glass-ceramics in dentistry. J. Mater. Sci. Mater. Med. **17**, 1037–1042 (2006)
Hornor, J.A.: Low-fusing temperature porcelain. US-Patent 5,552,350, 1996
Kamiya, T., Inoue, M., Inada, H.: Dental porcelain material preventing yellow coloration and method for producing same. US-Patent 5,466,285, 1995
Kim, S., Lee, Y., Lim, B., Rhee, S., Yang, H.: Metameric effect between dental ceramics and ceramics repairing resin composite. Dent. Mater. **23**(3), 374–379 (2007)
Lee, Y.K., Powers, J.M.: Metameric effect between resin composite and dentin. Dent. Mater. **21**, 971–976 (2005)
MacDowell, J.F., Beall, G.H.: Immiscibility and crystallization in Al_2O_3–SiO_2 glasses. J Am Ceram Soc **52**(1), 17–25 (1969)
Mazzi, F., Galli, E., Gottardi, G.: The crystal structure of tetragonal leucite. Am. Miner. **61**, 108–115 (1976)
Neuber, J.: Dental crowns. US-Patent 290019 (A1), 2006
O'brien, J.: Methods for determining the proper coloring for a tooth replica. US-Patent 4654794, 1987
Prasad, A.: Metallization of ceramic restorations. US-Patent 6,627,248, 2003
Rouf, M.A., Hermansson, L., Carlsson, R.: Crystallization of glasses in the primary plan field of leucite in the K_2O–Al_2O_3–SiO_2 system. Trans. J. Br. Ceram. Soc. **77**, 36–39 (1978)
Sago, S., Sakakibara, T.: Dental ceramics. JP-Patent 190215 (A), (2007)
Salomonson, J., Yanez, J.: Method for making ceramic artificial dental bridges. US-Patent 7,600,398, 2006
Sean, R.: Method for forming polychromatic pressable ceramics dental restoration. US-Patent 70191 A1, 2008
Sekino, M., Shioda, M.: Dental ceramics composition. JP-Patent 137847 (A), 2009
Shioda, M., Sekino M.: Dental ceramics material composition. JP-Patent 308415 (A), 2007
Shirakura, A., Lee, H., Geminiani, A., Ercoli, C., Feng, C.: The influence of veneering ceramics thickness of all-ceramic and metal ceramic crowns on failure resistance after cyclic loading. J. Prosthet. Dent. **101**, 119–127 (2009)
Tagami, J., Ikeda, M.: Dental restoration and method for producing the same, and ceramics paste for dental restoration. US-Patent 215010 (A1), 2009
Taylor, D.: Thermal expansion data XV. Complex oxides with the leucite structure and frameworks based on six-membered rings of tetrahedra. Journal **90**(6), 197–204 (1991)
Taylor, D., Henderson, C.M.B.: The thermal expansion of the leucite group of minerals. Am. Miner. **53**, 1476–1489 (1968)
Yi, Z., Rao, P., Lu, M., Wu, J.: Mechanical properties of dental ceramics with different leucite particle sizes. J. Am. Ceram. Soc. **91**(2), 527–534 (2008)
Yu, H.Y., Ca, Z.B., Ren, P.D., Zhu, M.H., Zhou, Z.R.: Friction and wear behaviour of dental feldspathic ceramics. Wear **26**(5–6), 611–621 (2006)

Chapter 11
Machinable Mica Dental Glass-Ceramics

11.1 Mica Glass-Ceramics

The metal-ceramic restoration is a well established solution for how to overcome the problem of the lack in strength and toughness of the highly aesthetic feldspathic dental ceramics. As shown in the previous chapter the primary consideration for the veneering ceramic is that it has a coefficient of thermal expansion that is appropriately matched to the coefficient of thermal expansion of the metal by the incorporation of leucite, that it bonds well to the metal substructure and provides aesthetic veneers able to mask the color of the underlying metal.

Despite considerable success of metal-ceramic restoration, it suffers from the disadvantage that the aesthetic will always be compromised to some degree due to the presence of the metal substructure, which prevents the natural processes of light transmission and scattering associated with the natural tooth.

An alternative approach is to do away with the metal substructure, but this requires a ceramic of considerably greater strength and toughness than it has as yet been possible to deliver with the leucite glass-ceramics. Also such a ceramic would benefit from the added structural strength that can be obtained from being bonded to the underlying tooth structure.

High strength ceramics such as alumina and zirconia lack the necessary translucency and therefore need to be veneered with a more aesthetic ceramic such as a feldspathic glass or leucite glass-ceramic. Alumina and zirconia core ceramics need to be optimized to prevent the serious obstacles to their ability to be bonded to the enamel and dentine. By contrast, silica-based glass-ceramics, such as mica glass-ceramics and lithium disilicate glass-ceramics, show considerable promise in this respect.

Mica glass-ceramics can potentially achieve strength in excess of 175 MPa and when used in combination with enamel and dentine bonding can to be used without the use of a metal support. The mica glass-ceramic system displays good aesthetic properties, mechanical strength, and machinability, which are successfully imparted by the crystallization of fluorophlogopite ($K_2Mg_6Al_2Si_6O_{20}F_4$) in the glass matrix. The high translucency produces a chameleon effect (produced as the restoration

E. El-Meliegy and R. van Noort, *Glasses and Glass Ceramics for Medical Applications*,
DOI 10.1007/978-1-4614-1228-1_11,

acquires color from adjacent teeth as a result of the translucency and color from the adjacent teeth). The unique microstructure of mica consists of randomly oriented small interlocking plate-like crystals. These crystals add strength and reinforcement, resulting in a flexural strength approximately twice that of the leucite glass-ceramics. This desirable combination of strength and beauty makes mica an excellent candidate as a dental ceramic.

A mica glass-ceramic can be made either through controlled crystallization of mica by bulk crystallization from the glass blocks or by the crystallization of mica crystals from glass frits by surface crystallization under controlled conditions. This chapter shows how it is possible to optimize the physical and mechanical properties of mica glass-ceramics and improve their mechanical machinability. A range of examples are provided that should appeal to dental materials scientists, dentists, and glass technologists. Methods to control the microstructure of mica fluorophlogopite, via controlling the chemical composition and the conditions of processing, are explored. In addition, we show how one might control ceramic properties such as mechanical strength, microhardness, chemical resistance, and aesthetics by the addition of various mineralizers.

11.2 Industrial Importance of Synthetic Mica

Mica is a strategic mineral upon which various industries are dependent. Micas are a family of minerals that have a similar crystallographic structure and are highly flexible and cleavable. Most natural micas may readily be split into a thickness of 1 mm or less to give a film of single crystal quality. The film is continuous in structure with a high dielectric strength that makes mica a very important electronic insulator.

Although natural mica is widely distributed, the principal commercial product (sheet and film) comes from only a few countries; primarily India and Brazil but this source is limited, necessitating the production of synthetic micas to meet demand. Synthetic micas are more desirable than natural micas because:

1. The supply of natural mica is a subject to interruptions or unavailability.
2. The sources of natural mica are rare and the quality and size are unpredictable.
3. The synthetic mica has higher purity, uniformity and is easily producible from abundant raw materials.
4. It is much easier to control the properties of synthetic micas such as thermal expansion coefficient, strength, machinability, microhardness and translucency through adjustment of the preparation conditions.

11.3 History of Synthetic Mica

The first successful synthesis of mica was made by Hautefeville et al. (1887). Iron mica was prepared by simple fusion and cooling of a composition using potassium silicofluoride (K_2SiF_6) as a source of potassium, silicon and fluorine. Mica was also prepared by the use of silicate-fluoride melts by Doelter in 1888. The synthesis of muscovite, biotite, phlogopite, or sodium mica was soon followed by using both natural materials and oxide chemicals.

Fluorine is sourced from sodium fluoride, potassium fluoride, and potassium silicofluoride. In the early days of mica production, the role of fluorine in the synthesis of mica was not fully understood. In addition, structural and crystal chemistry had not been sufficiently developed at this time to allow conclusions in this field. Nagai and Noda et al. (1938) in Japan started to prepare synthetic fluorine micas as did Reichmann et al. (1942) in Germany and they were able to synthesize fluorophlogopite and other micas with excellent quality. Reichmann et al. apparently made one single crystal, which may have been 30 cm long.

Eitel et al. (1953) made a systematic and fundamental study of the synthesis of micas. Various fluorides were studied to determine which one might be suitable for the synthesis of fluorophlogopite mica ($K_2Mg_6Al_2Si_6O_{20}F_4$). Fluorides of K_2SiF_6, MgF_2, K_2AlF_5, and K_3AlF_6 were found to be favored and in that order. KF, $KMgF_3$, AlF_3, $KAlF_4$, and $MgSiF_6$–$6H_2O$ are less suitable for the purpose.

Solid-state reaction studies were made on a large number of binary partial systems of the mica batch, including: $MgF_2 + MgO$, $3MgF_2 + Al_2O_3$, $2MgF_2 + SiO_2$, $3MgF_2 + KAlSi_3O_8$, $K_2SiF_6 + 3MgO$, $K_2SiF_6 + Al_2O_3$, $K_3AlF_6 + 3MgO$, and $K_3AlF_6 + 3SiO_2$. Solid-state reactions succeeded in preparing forms of fluorophlogopite at temperatures as low as 750°C from glass precursors. For best results, however, mica batches should be melted at 1,000–1,300°C in closed containers, using anhydrous batch materials to minimize hydrolysis of the fluorides and loss of HF.

The development of the first dental mica glass-ceramics was based on the strengthening of glass with fluorophlogopite mica. The lost wax system was used to produce a glass casting of the dental restoration. The casting was then subjected to a heat treatment during which fluormica crystals were formed to increase the strength and machinability of the glass-ceramic.

Glass-ceramics comprising a crystalline phase that belongs to the mica family such as tetrasilicic fluormicas or fluorophlogopite micas are known to exhibit excellent machinability. The glass-ceramics are useful in the CAD-CAM fabrication of single and multi-unit dental restorations including bridges, space maintainers, tooth replacement, splints, crowns, partial crowns, dentures, posts, teeth, jackets, inlays, onlays, facing, veneers, facets, implants, abutments, cylinders, and connectors. The micaceous glass-ceramics can also colored to adequately match the shades of the surrounding natural teeth.

11.4 Crystalline Structure of Mica

The micas constitute a family of silicate minerals having a unique two-dimensional sheet structure. Most naturally occurring micas are hydroxyl silicates, whereas micas produced synthetically have commonly involved replacing the hydroxyl groups within the crystal structure with fluorine. These synthetic micas are termed fluormicas.

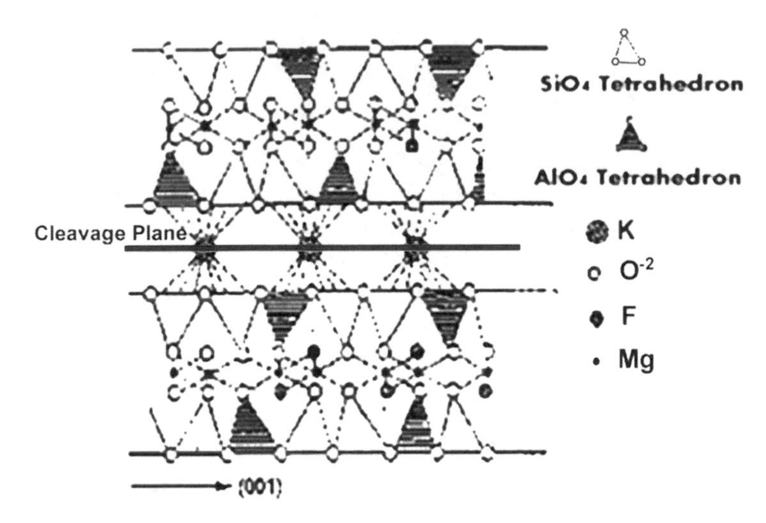

Fig. 11.1 The structure of mica, consisting of an octahedral layer sandwiched between two tetrahedral layers after Chen et al. (1998)

The classic crystal structure of fluormica has been defined within the generalized structural formula $X_{0.5-1}Y_{2-3}Z_4O_{10}$ $(OH_3F)_2$. The *X* represents cations, which are relatively large in size, e.g., 0.9–1.6 Å radius such as potassium, *Y* signifies somewhat smaller cations, e.g., 0.5–0.9 Å radius for example Mg, and *Z* describes small cations, e.g., 0.25–0.5 Å radius such as Si, which coordinate to four oxygen. The *X* cations are in dodecahedral coordination and the *Y* cations in octahedral coordination.

The fundamental unit of the mica structure is the Z_2O_5 hexagonal sheet resulting from the fact that each ZO_4 tetrahedron shares three of its corners with other tetrahedra. Thus, the structure of fluormica is similar to that of naturally occurring mica in that two Z_2O_5 sheets, each having apical oxygen and associated interstitial fluoride ions directed toward each other, are bonded by the *Y* cations.

The mica layer so-formed has been demonstrated to be a 2–1 layer since it is composed of two tetrahedral sheets with one octahedral sheet. The individual mica layers are bonded to each other through the relatively large *X* cations in the so-called interlayer sites. The *X* cations are usually potassium but can be other large alkali metal and alkaline earth metal cations such as Na^+, Sr^{+2}, Ba^{+2}, Rb^+ and Cs^+. The *Y* cations can be Li^+, Mg^{+2}, and Al^{+3}. The *Z* cations will be Si^{+4}, Al^{+3}, and, perhaps B^{+3} (Beall et al. 1976).

The structure of mica is composed of an octahedral layer sandwiched between two tetrahedral layers pointing toward each other (Fig. 11.1). The octahedral layer consists of closely packed oxygen, hydroxyl, and/or fluorine in which cations of radii from 0.5 to 0.8 Å are present. The tetrahedral layer consists of silicon–oxygen tetrahedrons linked together in a hexagonal network. The three layer composite sheet has a net negative charge. The excess charge is balanced by the uptake of cations between the composite sheets in 12-fold coordination. The interlayer cations are weakly bonded giving rise to a perfect set of cleavage.

The structural transformations of mica are followed by a change in the chemical and thermal properties. Also, the isomorphous substitution has been extended to include the different cations in the structure. The isomorphous substitutions include the tetrahedral layer where silicon is found in the structure. The substitution follows the solid solution chemistry of mica expressed as $K_1 - x Mg_3 - y Aly((Al,B)1 - z Si_3 + z O_{10} + w)F2\text{-}w$ assuming the sum of (O + F) anions = 12.

Continuous extension of this linkage creates a hexagonal network within each tetrahedrally coordinated plane. The two tetrahedral sheets are strongly bonded together by Mg ions, which are in sixfold or octahedral coordination. The F ions, although lying in the plane of the apical O ions, are bonded to the Mg cations only. Thus the anions forming the octahedral site are two F ions and two O ions. The K ions, in 12-fold coordination, bond adjacent double sheets together. The cleavage plane, characteristic of mica minerals, lies between sets of double sheets. The cations in tetrahedral coordination are usually either Si or a combination of Si and Al. Micas containing only Si in the tetrahedral sites are called tetrasilicic; those containing an average Si:Al ratio of 3:1 in fourfold coordination, trisilicic; and those with equal numbers of Si and Al in YO, coordination, disilicic.

The octahedral sites formed by the two F and four O ions are occupied by cations having ionic radii of 0.6–0.9 A, most commonly Mg^{2+}, Al^{3+}, Mn^{2+}, Ti^{4+}, and Li^{+}. Two distinct groups of mica compounds (as determined by the number of filled octahedral positions) are known, i.e., those having 2/3 of the available sites filled (dioctahedral) and those having all octahedral positions filled (trioctahedral).

11.5 Structure of Fluorophlogopite

The formula of potassium fluorophlogopite is $KMg_3AlSi_3O_{10}F_2$ (Or the formula is doubled $K_2Mg_6Al_2Si_6O_{20}F_4$). The doubled formula represents the actual unit cell and indicates two silica sheets per layer of mica. Mica can also be distinguished according to the content of silica into disilicic ($BaMg_3Al_2Si_2O_{10}F_2$), trisilicic fluorophlogopite ($KMg_3AlSi_3O_{10}F_2$), and tetrasilicic ($KMg_2LiSi_4O_{10}F_2$). The fluorine micas are synthesized at ordinary atmospheric pressure. The process used is known as dry synthesis. Hydroxymica, on the contrary, is prepared using a hydrothermal technique.

The structure of the tetrasilicic synthetic mica ($K_2Mg_5Si_8O_{20}F_4$) is different from that of trisilicic fluorophlogopite ($KMg_3AlSi_3O_{10}F_2$) in terms of the amount of Al and Si substitution in the tetrahedral level. Basically it consists of two sheets of SiO_4 tetrahedra with their vertices pointing inward. These tetrahedra are linked in one plane by sharing the three oxygen atoms which form the base of each tetrahedron, each anion being shared by two tetrahedra as shown in Fig. 11.2.

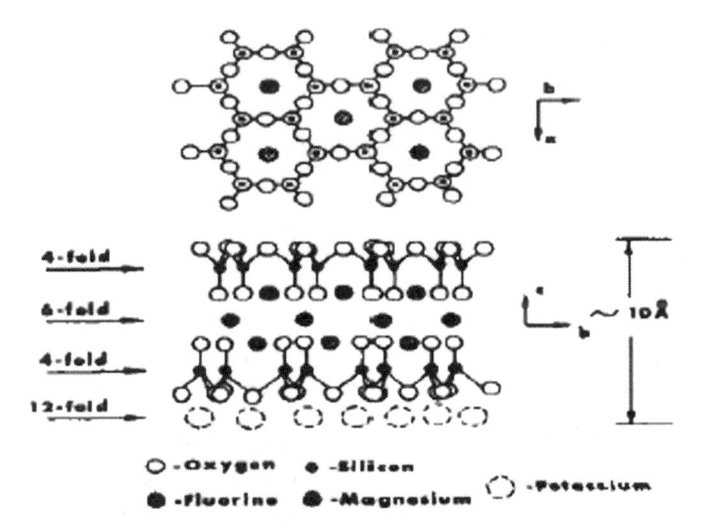

Fig. 11.2 Structure of tetrasilicic mica after Daniels and Moore (1975a, b)

11.6 Chemical Reactions of Mica and Mica Related Phase Diagrams

The reaction between MgF_2 and Al_2O_3 in a closed system at 1,100°C yields HF and $MgAl_2O_4$ spinel. In open containers at 1,000–1,300°C, MgO reacts with SiO_2 to form forsterite or with MgF_2 and silica form norbergite $Mg_2SiO_4{\cdot}MgF_2$.

The most important reaction in the synthesis of fluorophlogopite is the reaction between potash feldspar and MgF_2. The mixture had all the constituents of fluorophlogopite and some fluorophlogopite is formed together with forsterite and residual MgF_2 at 1,000°C.

11.6.1 MgO–MgF$_2$–SiO$_2$ System

In the system MgO–MgF_2–SiO_2 (Fig. 11.3), only the humite minerals with norbergite, $Mg_3(SiO_4)(F,OH)_2$, and chondrodite $(Mg,Fe,Ti)_5(SiO_4)_2(F,OH,O)_2$ crystallize by the solid-state reactions in the system MgO–MgF_2–SiO_2 and clinohumite $(Mg,Fe)_9(SiO_4)_4(F,OH)_2$ and humite $(Mg,Fe)_7(SiO_4)_3(F,OH)_2$ will crystallize under special conditions. Only norbergite, chondrodite, and the expected cristobalite, clinoenstatite, and forsterite were designated as having the primary field of crystallization.

Norbergite melts with a primary crystallization of chondrodite (stable), or forsterite (metastable) at 1,345°C. Chondrodite melts incongruently with crystallization

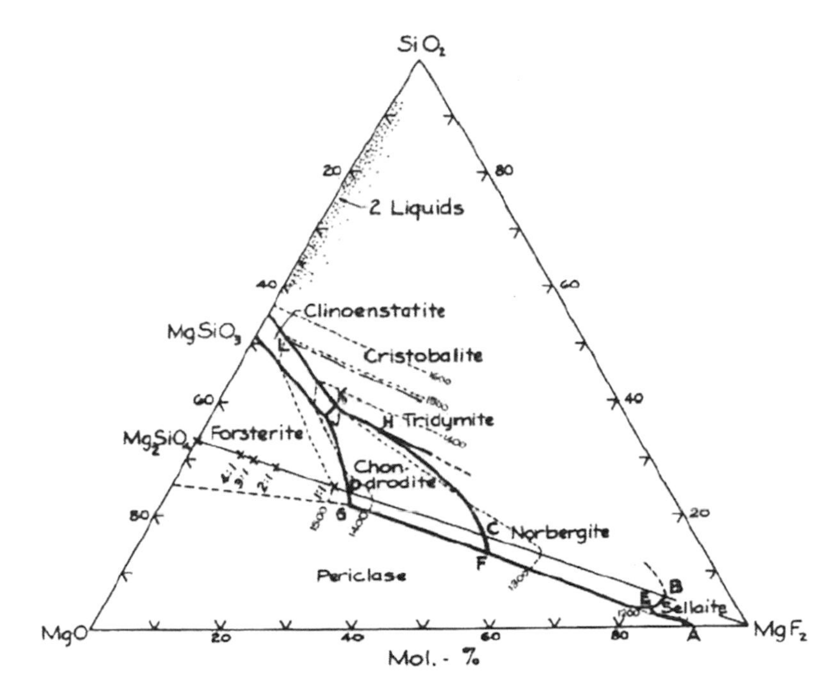

Fig. 11.3 System $MgO–SiO_2–MgF_2$, showing compatibility triangles up to 1,300°C after Hinz and Kunth (1960)

of forsterite at 1,450°C. Clinohumite breaks down at 1,380°C and yields a mixture of forsterite and chondrodite. Humite has been observed only occasionally and even then in a subsolidus or solid-state reactions, it is not a primary phase from any liquid.

The equilibrium temperature for the reaction Clinohumite + Humite = Chondrodite, could not be determined. Despite its metastability, forsterite rather than chondrodite is usually formed from the fusion of norbergite. Cristobalite is the predominant silica phase in the base system $MgO–MgF_2–SiO_2$ whereas tridymite is the predominant phase in the presence of Li_2O, Na_2O and/or K_2O.

Noda (1955) placed the liquidus at 1,361 °C and stated that specimens quenched from 1,363°C contained needle like crystals of forsterite and no mica, while specimens quenched at 1,359°C, i.e., just below the liquidus temperature, were composed of mica with small amount of forsterite. The melting point of fluorophlogopite is 1,375 ± 5°C.

The fluorophlogopite composition is located either just within the field of stability of forsterite or is right on the boundary line between its own field and that of forsterite. Probably, fluorophlogopite actually melts incongruently within 1–5°C, with forsterite being the primary phase to separate from a melt of fluorophlogopite composition (Fig. 11.4).

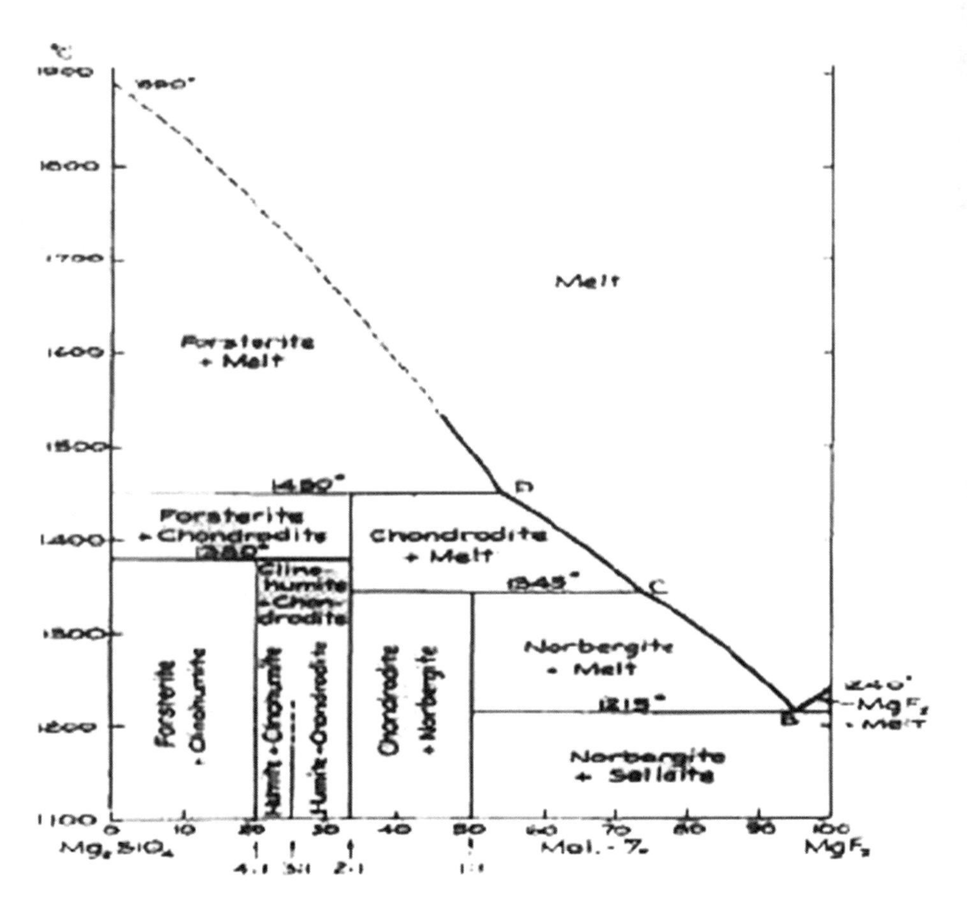

Fig. 11.4 MgF_2–Mg_2SiO_4 system after Hinz and Kunth (1960)

11.7 Chemical Compositions of Mica Glass-Ceramics

In multiphase glass-ceramics, each phase contributes distinctive characteristics. The glass-ceramic containing synthetic fluormica can be derived through the crystallization of glasses.

Useful medical mica compositions should include SiO_2, MgO, K_2O, F, Al_2O_3, ZrO_2 and BaO. On the contrary, B_2O_3 is an important ingredient in the crystallization of fluormica glass-ceramic exhibiting good mechanical machinability. However, its presence is undesirable in some cases due to its effect on reducing the coefficient of thermal expansion and refractoriness. Other minor additions of oxides to the base glass composition, generally in amounts no more than 10% in total, such as P_2O_5, CaO, CdO, GeO_2, FeO and ZnO that may act to improve the melting and forming behavior or the physical properties of the final product.

An example of a glass-ceramic composition that is particularly suitable for the fabrication of dental crowns is the tetrasilicic fluormica glass-ceramic, containing about 2% of BaO, which is of beneficial effect on the hydration stability of the glass-ceramics and getting rid of surface porosity formation. On the contrary, the lower content of BaO adversely affects the translucency.

Ooishi et al. (1991) developed a process for the preparation of 30–60 vol% fluorophlogopite-containing machinable ceramics with fine crystals evenly dispersed in a vitreous matrix through a solid-phase reaction.

Kasuga et al. (1993) prepared a glass-ceramic with excellent mechanical strength, machinability, and chemical durability, suitable for the production of an artificial dental crown. The glass-ceramic contains 2–17% CaO, 0–5% K_2O, and 0–4% Na_2O, the total content of K_2O and Na_2O being 0.5–7%, 15–35% MgO, 30–49% SiO_2, 2–15% ZrO_2, and not more than 14% of fluorine. CaO has the effect of precipitating mica crystals, improves the machinability, and lowers the viscosity of the glass. However, the resultant glass-ceramic is poor in machinability at low CaO content.

The content of K_2O and Na_2O affects the chemical solubility by precipitating a fine mica crystal by heat treatment depending upon the ratio of glass matrix/mica. With a decrease in the residual glass, the chemical solubility is improved. When, the K_2O content exceeded 5 wt%, devitrification tends to occur, while if Na_2O content exceeds 4%, a sodium mica crystal precipitates resulting in an undesirable decrease in mechanical strength.

MgO is one of the main components in a mica crystal. It is difficult to precipitate a crystalline phase when the MgO content is less than 15 wt%. The glass tends to devitrify when the SiO_2 content is less than 30 wt% and it is difficult to obtain a homogeneous glass when the SiO_2 content exceeds 50 wt%.

Al_2O_3 is a main component affecting the stability of developed mica crystals and improving chemical durability, where the glass tends to devitrify when the Al_2O_3 content is less than 5 wt% and the viscosity increases when the Al_2O_3 content exceeds 30 wt%.

ZrO_2 can act as a nucleating agent and improve the mechanical strength of the glass. It produces an effect as a nucleating agent even at a low content of less than 2%, and fine mica crystals can be precipitated. Fluorine is useful in controlling the precipitation of mica crystals, playing a big role in the phase separation process and it is difficult to precipitate mica crystals in the glass at fluorine content lower than 1.5%.

In addition to the above essential components, other oxide components such as SrO, TiO_2, Nb_2O_5, Ta_2O_5 and Y_2O_3 can be used so long as the total content of these oxides is kept to less than 5 wt% to avoid devitrification.

In addition, other crystalline phases may evolve either desirable or undesirable properties for medical applications if the starting chemical composition is not carefully managed, as there is a wide chemical substitution of various elements in the structure of mica. These phases may include enstatite ($MgO{\cdot}SiO_2$), an akermanite ($2CaO{\cdot}MgO{\cdot}2SiO_2$), a diopside ($CaO{\cdot}MgO{\cdot}2SiO_2$), an anorthite ($CaO{\cdot}Al_2O_3{\cdot}2SiO_2$), richterite ($Na_2CaMg_5Si_8O_{22}F_2$), and a forsterite ($2MgO{\cdot}SiO_2$). These phases can improve the mechanical strength. A glass-ceramic thus obtained is not only excellent in terms of mechanical strength and machinability but also excellent in chemical durability and could be used to produce a dental crown, as shown in Table 11.1.

Table 11.1 Four examples of multiple-phase glass-ceramics given by Kasuga et al. (1993)

CaO	7.5	16.5	2.2	3.4
K_2O	0.3	4.8	1.4	0.7
Na_2O	0.5	0.2	0.2	2.2
MgO	34.7	20.5	17.2	19.2
SiO_2	30.7	43	49	41
Al_2O_3	12.3	5.3	22.9	21.6
ZrO_2	6	4	2.5	5.6
F	8	5.7	3.4	7.5
Temp. (°C)	850	1,050	950	950
Time (h)	10	10	2	2
Phases	MZE	MZED	MZA	MZA
Bend strength (Mpa)	250	350	320	300
Machinability	Excellent	Machinable	Excellent	Excellent

M mica, *Z* zirconia, *E* enstatite, *D* diopside, *A* akermanite

11.8 Development of Mica Glass-Ceramics Microstructures

The crystalline microstructure is defined by the crystal size and morphology and the textural relationship among the crystals and glass. The microstructure represents the key to many mechanical and optical properties, opacity, strength, fracture toughness, and machinability. These microstructures of glass-ceramics can be quite complex and distinct.

The microstructures of mica either fluorophlogopite trisilicic ($KMg_3AlSi_3O_{10}F_3$) or tetrasilicic mica ($KMg_{2.5}Si_4O_{10}F_2$) are known to develop through nucleation and crystallization. Fine crystals evenly dispersed in the glassy matrix is the main driving force for producing a glass-ceramic with good machinability, mechanical, chemical and physical properties. The development of the mica glass-ceramics microstructure by the glass crystallization is complex, but it can give rise to promising properties for medical applications.

On cooling of the glass melt, some mica glasses tend to undergo spontaneous liquid–liquid phase separation, giving rise to a white opal color. The opalization is due to the formation of fluorine-rich droplets. On seeking to produce a smooth surface layer using tetrasilicic mica glass-ceramics, what is achieved instead is a rough or porous surface film. The phenomenon of surface roughening was not limited to veneered constructs, but in mica objects as well by processing through the standard firing cycle. Obviously, surface porosity of any kind would be unacceptable in dental constructs for aesthetics and hygiene purposes. The primary reason of this surface effect is believed to be moisture adsorbed by the tetrasilicic fluormica, which is violently driven off during the firing to disrupt the surface. While it is possible to avoid the water adsorption problem, e.g., by veneering the construct immediately after polishing and/or by using a low temperature firing treatment. The problem can be resolved by the addition of controlled amounts of BaO, which has been found to eliminate the surface disruption during firing, reduces the extent of water adsorption, improve the strength, chemical durability and the translucency of the glass-ceramics

Table 11.2 Compositions of tetrasilicic mica after Grossman et al. (1976)

Constituents	Amount (wt%)			
	I	II	III	IV
K_3O	11.0	13.5	15.5	15.1
MgF_2	10.6	10.4	10.2	10.1
MgO	16.4	13.5	13.2	12.3
SiO_2	62.0	60.5	59.1	58.8
AS_2O_5		2.0	2.0	1.9
ZrO_2				1.9

(Grossman et al. 1987). Nevertheless, this surface finish problem severely inhibit widespread use of micaceous materials as dental restorations considering that the driving force for all-ceramic restorations is their aesthetics superiority over metal-ceramic restorations. These micaceous dental ceramic are commercially available for use in CAD/CAM. For the above reason, the trisilicic fluorophlogopite micas are suitable for the use in dental constructs.

11.9 The Crystallization of Tetrasilicic Mica

The crystallization sequence of a tetrasilicic mica composition is quite different from that of the trisilicic fluorophlogopite mica. As far as can be ascertained tetrasilicic mica crystals nucleate homogeneously from the glass without any of the complex phase transformations observed in the trisilicic mica.

Tetrasilicic mica glass-ceramics may be produced from glasses based on the simple quaternary system K_2O–MgF_2–MgO–SiO_2. Extensive substitutions for the K^+ ions may be made by using divalent cations such as Sr^{2+} and Ba^{2+}. The tetrasilicic fluormica composition ($K_2Mg_5Si_8O_{20}F_4$) crystallizes in a 2-stage crystallization mechanism in compositions slightly deficient in F. Some compositions of tetrasilicic mica are shown in Table 11.2.

A phase separation leads to the opalization and appearance of quasi-spherical mica grains at temperatures as low as 625°C. The opalization at temperatures above the crystallization range is of a critical significance to the mica crystallization. The phase separation is thus a significant factor in influencing the nucleation and crystal growth rates. On cooling of the glass melt, some mica glasses tend to undergo spontaneous liquid–liquid phase separation, giving rise to white opal color. The opalization is due to the formation of fluorine-rich droplets.

The crystal growth is superimposed upon a phase-separated glass. The mica grains recrystallize at high temperature by grouping the fine scale spherical mica grains into larger size of about 1–2 μm. Increasing the F content to the stoichiometric composition (composition equivalent to $K_2Mg_5Si_8O_{20}F_4$), lowers the transition temperature. Also minor additions of various compatible metal oxides, such as Li_2O, K_2O, Al_2O_3, SrO and Fe_2O_3 are tolerated up to a total of about 15% wt% and can be substituted into the mica structure. P_2O_5 and B_2O_3 can be incorporated into the residual glassy matrix to lower the maturing temperature and alter the coefficient

Table 11.3 Physical properties of tetrasilicic mica after Grossman et al. (1976)

	Composition	
Property	I[a]	II[b]
Density (g/cm²)	2.67	2.61
Porosity	0	0
Mechanical properties		
Modulus of rupture (psi)	22,900	10,200
Compressive strength (psi)	75,000	65,000
Modulus of elasticity (psi)	9.88×10^4	9.44×10^4
Shear modulus (psi)	3.94×10^4	3.74×10^4
Poisson's ratio	0.25	0.26
Thermal properties		
Chemical durability (wt loss in mg/cm_2)		
50% HCI for 24 h at 95°C	0.69	5.3
N/50 Na_2CO_3 for 6 h at 95°C	1.0	0.09

[a]Heated to 1,150°C for 4 h
[b]Heated to 1,100°C for 4 h

of thermal expansion. The usual annealing times of the glass involves periods of about 2 h while a very long annealing schedule does not improve the crystallization or internal microstructure of the final product.

The mechanical properties of the mica-containing glass-ceramic, especially the machinability depend on the degree of interlocking of the mica crystals and the vol% of the mica crystals in the glassy matrix. The physical properties of tetrasilicic mica are shown in Table 11.3.

11.10 The Crystallization of Fluorophlogopite (Trisilicic Mica) Glass-Ceramics

Fundamentally, the desirable crystallization of mica involves nucleation and controlled growth of crystals at or near the liquidus temperature to build an evenly dispersed fine-grained microstructure. The chemical composition of the glass plays a major role in the crystallization of mica. Although opalization can be observed in the starting glasses that make fluorine micas, fine crystallites are precipitated at temperatures above the annealing point during or immediately after the opalization and provide the available nuclei demanded for initiating the crystallization.

On heating fluorophlogopite mica based glass above the annealing point, MgF_2 crystals precipitate from fluorine-rich droplets and a series of phase transformations evolve. The eventual microstructure depends on the composition of the fluorophlogopite crystals in the glass matrix. Very fine crystals have been identified by XRD to be norbergite ($Mg_2SiO_4MgF_2$). A combination of MgF_2 and some mullite crystals with a potassium-rich glassy phase can lead to the development of a characteristically two-dimensional fluorophlogopite. The number of mica crystals increases through growth on norbergite crystallites on further heating until all of the norbergite disappears. On heat treatment at 800°C/4 h and at 950°C/2 h, the final microstructures consists of small fluorophlogopite crystals dispersed in the glassy matrix, together with a minor amount of secondary mullite or forsterite. Mullite crystals are found to disappear at temperature greater than 850°C.

A very good example is a fluorophlogopite based on a talc–feldspar mixture utilized for the processing of dental porcelain (Mustafa 2001). A uniform microstructure with fine equigranular fluorophlogopite grains is developed by heat treating the glass at a temperature of 950°C. Fluorophlogopite grains of 0.2 μm in size are well distributed in the glassy phase. The crystallization of fluorophlogopite occurs at 910°C.

The content of fluorine plays an essential role in the crystallization of mica fluorophlogopite. The course of the chemical reaction completely changes on the addition of MgF_2 to talc and feldspar mixture. Fluorophlogopite forms instead of a two phase mixture of forsterite and enstatite.

On firing at 1,050°C or higher, small amount of forsterite appear, which are associated with mica grain growth. The formation of forsterite is related to the decomposition of fluorophlogopite. The use of mineralizers does not prevent the formation of forsterite. This fact indicates that the forsterite is formed as a result of the partial decomposition of fluorophlogopite. The decomposition is related to fluorine volatilization at higher temperatures. The amount of fluorine not only affects the temperature of crystallization but also affects the phase transformation and fluorophlogopite decomposition.

The intensities of different fluorophlogopite XRD peaks may be slightly different from that recorded in the XRD card. The intensities of peaks with hkl (003) and (001), corresponding to 65 and 100% respectively, can display different intensity values. The difference is possibly due to a kind of preferred orientation of the mica grains.

A simple addition of MgF_2 and B_2O_3 to the talc–feldspar mixture reduces the temperature of melting to 1,200°C and enhances the crystallization of fluorophlogopite such that it will form at the lowest possible temperature. Studying the thermal behavior is necessary to avoid fluorine volatilization and weight loss as well as to control grain growth. Two grain morphologies appear in the glass-ceramic when fired at 950°C/2 h, including tabular and prismatic crystals distributed in the glassy phase with different orientations. The size of grains ranges from less than 0.5 μm in length for prisms and less than 0.3 μm in size for tabular grains as shown in Fig. 11.5. Prismatic grains are not well developed.

The first crystallization evolves at 645°C. The thermal expansion of mica glass-ceramics, produced by melting a mixture simply made by the addition of MgF_2 and

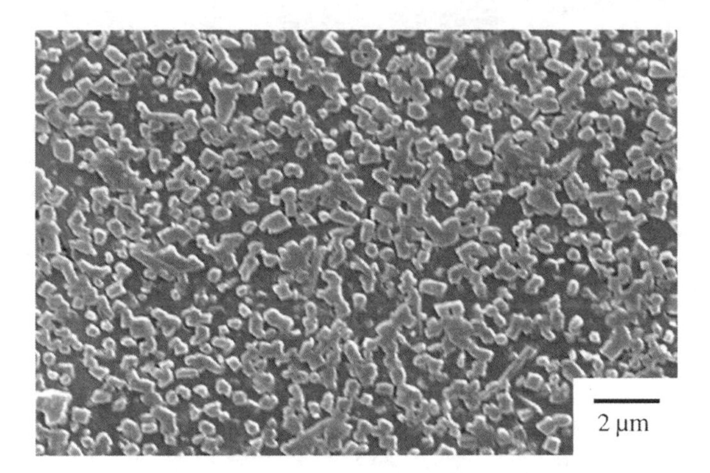

Fig. 11.5 Fluorophlogopite mica made from a feldspar–talc mixture, Mustafa (2001)

B_2O_3 to talc–feldspar mixture, can be adjusted to yield a uniform microstructure and tailored thermal expansion changes. Linear thermal expansion occurs up to 540°C, corresponding to the temperature at which fluorine ions start to be ordered to act as a nucleation agent for the formation of octahedra. The rate of expansion increases gradually up to 645°C corresponding to the temperature of the first crystallization. A linear expansion curve is observed between 645 and 780°C. The coefficient of thermal expansion is 107×10^{-7}/°C in the temperature range from 20 to 780°C.

A uniform microstructure, with fine, equigranular and uniformly distributed fluorophlogopite grains in the glassy matrix is formed at 950°C. The morphological and microstructural properties of fluorophlogopite mica can be controlled by adjusting the rate of firing to 5°C/min and a heat treatment at 700°C/2 h before crystallization. The formation of fluorophlogopite is preceded by the formation of norbergite as indicated by the results of DTA at 750°C. The norbergite phase nucleates through quenching of the silicate melt, where the possibility for nucleus formation increases rapidly with supercooling. The nucleation process is also enhanced by thermal treatment at 700°C/2 h in addition to a low rate of firing. The crystallization of perfect fluorophlogopite will occur at 910°C as detected by DTA and confirmed by XRD and SEM at 950°C/2 h.

11.11 Scientific and Technical Problems Encountered in Synthetic Mica Glass-Ceramics for Dental Applications

Despite good mechanical strength, good machinability, and excellent aesthetics, the application of mica glass-ceramics in dentistry still faces many problems. A deficiency of fluorine during glass making results in the decomposition of mica into forsterite (Mg_2SiO_4) and norbergite (Mg,SiO,MgF_2). Grossman has reported that enstatite ($MgSiO_3$) develops at ~980°C in the crystallization of tetrasilicic mica glass-ceramic of a composition deficient in F.

Further Reading

Abert, C., Beleites, E., Carl, G., Grosse, S., Gudziol, H., Hoeland, W., Hopp, M., Jacobi, R., Jungto, H., Knak, G., Kreisel, L., Musil, R., Naumann, K., Vogel, F., Vogel, W.: Micaceous-cordierite-glass ceramic. US Patent 4,789,649, 1988

Comeforo, J.E., Hatch, R.A., Humphrey, R.A., Eitel, W.: Synthetic mica investigations: I, a hot-pressed machinable ceramic dielectric. J. Am. Ceram. Soc. **36**(9), 286–294 (1953)

Comyns, A.E.: Fluoride Glasses: Critical Reports on Applied Chemistry, vol. 27. Society of Chemical Industry, John Willey & Sons, Chickester, New York, Toronto, Singapore (1989)

Eitel, W., Hatch, R.A., Denny, M.V.: Synthetic mica investigations: 11, role of fluorides in mica batch reactions. J Am Cream. Soc. **36**(10), 241–248 (1983)

Grossman, D.G.: Machinable glass-ceramics based on tetrasilicic mica. J. Am. Ceram. Soc. **55**, 446–449 (1972)

Hakamatsuka, Y., Watanabe, K.: Glass ceramic dental crown and method of manufacturing the same. US Patent 4,799,887, 1989

Hoda, S.N., Beal, G.H.: Alkaline earth mica glass-ceramics, advances in ceramics: nucleation and crystallization in glasses. Am. Ceram. Soc. **287–299**(4) (1982)

Höland, W., Vogel, W., Naumann, K., Jummel, J.: Interface reactions between machinable bioactive glass-ceramics and bone. J. Biomed. Mater. Res. **19**, 303–312 (1985)

Sarver, J.V., Hummel, F.A.: Stability relations of magnesium metasilicate polymorphs. J. Am. Ceram. Soc. **55**(4), 152–157 (1962)

Schneider, I., George J., Taub, L.: Method of fabrication of translucent dental restorations without opacious substructures. US Patent 6,033,222, 2000

Steidl, J., Assmann, S.: Ceramic dental restoration. US Patent 6,342,302, 2002

Wu, J.M, Cannon, W.R., Panzera, C.: Castable glass-ceramic composition useful as dental restorative. US Patent 4,515,634, 1985

References

Beall, G.H.: Mica-spodumene glass-ceramic articles. US Patent 3,997,352, 1976

Chen, X., Hench, L., Greenspan, D., Zhong, J., Zhang, X.: Investigation on phase separation, nucleation and crystallization in bioactive glass-ceramics containing fluorophlogopite and fluoroapatite. Ceram. Int. **24**(5), 401–410 (1998)

Daniels, W.H., Moore, R.E.: Crystallization of a tetrasilicic fluormica glass. J. Am. Ceram. Soc. **58**(5–6), 217–221 (1975)

Eitel, W., Hatch, R.A., Denny, M.: Synthetic mica investigations: II, role of fluorides in mica batch reactions. J. Am. Ceram. Soc. **36**(10), 341–348 (1953)

Grossman, D.G.: Spontaneously-formed fluormica glass-ceramics. US Patent 3,985,531, 1976

Hautefeville, P., Compt. rend. 104, 508 (1887)

Hinz, W., Kunth, P.O.: Phase equilibrium data for the system MgO–MgF_2–SiO_2. Am. Miner. **45**, 1198–1210 (1960)

Kasuga, T., Kasuga, T.: Glass-ceramic and artificial dental crown formed therefrom. US Patent 5,246,889, 1993

Mustafa, E.: Fluorophlogopite porcelain based on talc feldspar mixture. Ceram. Int. **27**(1), 9–14 (2001)

Noda, T.: Synthetic mica research in Japan. J. Am. Ceram. Soc. **38**(4), 147–152 (1955)

Ooishi, T., Matsumoto, A.: Production process of machinable ceramics. US Paent 5,026,412, 1991

Reichmann, R., and Middle, V.: Production of synthetic mica, A Report on the synthetic mica at Siemens-Schuckert, issued in 1942

Chapter 12
Lithium Disilicate Glass Ceramics

12.1 Lithium Disilicate Glass-Ceramics

In this chapter, the best way to crystallize and process lithium disilicate glass ceramics for dental applications is discussed. In addition, the best way to modify the chemical composition and optimize the microstructure to achieve lithium metasilicate or lithium disilicate is assessed.
The lithium disilicate glass-ceramic exhibits a high mechanical strength that may reach up to 400 MPa, which compares favorably with that of many other ceramic used in dentistry. In addition, it is readily prepared from the glass melt by casting and cooling in commercially available dental laboratory investment molds and can be made into blocks for CAD–CAM processing.

Lithium disilicate glass ceramics are currently used in the fabrication of single and multiunit dental restorations including bridges, space maintainers, tooth replacement appliances, splints, crowns, partial crowns, dentures, posts, teeth, jackets, inlays, onlays, facing, veneers, facets, implants, abutments, cylinders, and connectors. Lithium disilicates have good pressability and processed by hot pressing or injection molding into refractory investment molds using the lost wax technique.

Other chemical components may be added to a certain extent in order to improve the chemical solubility such as Al_2O_3 or ZnO. The resultant lithium disilicate in the system Li_2O–Al_2O3–SiO_2 is generally superior to the simple lithium oxide–silicon dioxide system in terms of chemical solubility. However, both systems exhibit such high melt viscosities and need such high processing temperatures of about 1,350–1,400°C to cast that it is energy wasteful and results in instability problems.

The crystallization of lithium disilicate within the Li_2O–CaO–Al_2O_3–SiO_2 system and the utilization of P_2O_5 as a nucleating agent are considered promising. The fracture toughness (K_{ic}) of lithium disilicate ranges from 3 to 4 MPa $m^{0.5}$, the biaxial strength is 250–350 MPa, and the three-point bending strength is around 300–400 MPa. CaO is an essential ingredient for making glass-ceramics with improved chemical solubility. The acid solubility can be reduced to considerably less than 2,000 μg/cm^2 after soaking in 4 vol% aqueous acetic acid solution according to ISO 6872-1995 after 16 h.

E. El-Meliegy and R. van Noort, *Glasses and Glass Ceramics for Medical Applications*,
DOI 10.1007/978-1-4614-1228-1_12,

The introduction of digital fabrication techniques offers reliability in making dental restorations with high strength and toughness. The lithium disilicate (Li_2O–$2SiO_2$) glass ceramic, which is highly esthetic with excellent mechanical strength and toughness, provides new options for improving the quality of dental restorations. Lithium disilicate offers a viable solution to fabricating ceramic restorations with high thermal shock resistance due to its low thermal expansion.

12.2 Advantages of Lithium Disilicate Glass Ceramics

One of the merits of lithium metasilicate glass ceramic is that it is easily machinable and can be converted by a further heat treatment into a lithium disilicate glass ceramic product having also excellent optical properties. The conversion from lithium metasilicate to a lithium disilicate glass ceramic is associated with a very small linear shrinkage of only about 0.2–0.3%, which is almost negligible in comparison to a linear shrinkage of up to 30% when sintering ceramics. The final lithium disilicate glass ceramic has not only excellent mechanical properties, such as high strength and toughness, but also displays other properties required for dental restorations. This can be achieved without the need for ZnO as a component, which would result in a disadvantageously strong opalescent effect. Using the right percentage of color components does not affect strength or toughness of the glass ceramic and the thermal behavior is not adversely impaired.

12.3 Crystallization of Lithium Disilicate Glass Ceramics

A lithium disilicate glass can be prepared from an appropriate glass batch by melting at about 1,200–1,450°C for 4 h, followed by quenching in cold water to form a frit. The glass powder can be sintered to crystallize using the surface crystallization mechanism. With the addition of a nucleating agent such as P_2O_5 both surface and volume crystallization take place. Thus, for the right composition that will bulk nucleate, the glass can be cast in the form of a block and then cooled to room temperature. The glass block is then annealed, nucleated, and exposed to volume crystallization at a temperature between 400 and 900°C. Depending on the composition, the nucleation step may be carried out in the range of about 450–700°C for a time of about 0.5–4 h and the crystal growth step may be carried out in the range of about 700–1,000°C for a time that can range from several minutes to several hours.

12.4 Crystalline Phase Development

Lithium disilicate crystallizes, in the Li_2O–SiO_2 system, from the glass via homogenous nucleation and crystal growth. The stoichiometric composition of lithium disilicate crystal phase (Li_2Si_2O5) melts congruently at 1,033°C. The structure of orthorhombic

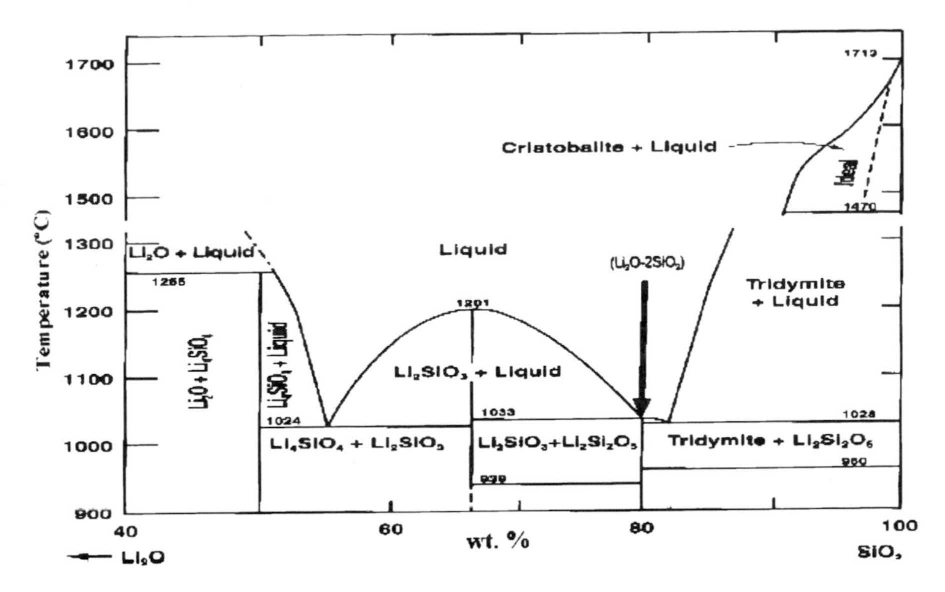

Fig. 12.1 The system Li_2O–SiO_2 (American Ceramic Society)

lithium disilicate crystals involves grooved sheets of $(Si_2O_5)^{2-}$ on the (010) plane resulting in excellent mechanical properties for the glass-ceramic. The phase diagram of the system Li_2O–SiO_2, after Kim and Sanders (1991), is shown in Fig. 12.1. The melting points of Li_2O and SiO_2 are 1,727 and 1,713°C, respectively. Different phases include $L_2S=Li_4SiO_4$; lithium metasilicate $=Li_2SiO_3$; lithium disilicate $=Li_2Si_2O_5$.

Lithium metasilicate crystallization was found to precede lithium disilicate crystallization as a metastable phase in some lithium silicate glasses and crystallizes at a lower temperature. Hench et al. (1971) determined that lithium metasilicate crystals form before lithium disilicate in two glass compositions of 30 and 33 mol% Li_2O, but the amount of lithium metasilicate was extremely limited and disappeared on further heat treatment at higher temperature.

As the Li_2O content increases over the stoichiometric content of lithium disilicate ($Li_2O{\cdot}2SiO_2$), lithium metasilicate (Li_2SiO_3) crystallizes and other crystalline phases can form depending on the starting glass composition. When the Li_2O is less than the stoichiometric content of the disilicate composition, metastable glass-in-glass phase separation occurs that influences crystal nucleation and crystal growth. Lithium disilicate and lithium metasilicate can crystallize by homogeneous bulk crystallization depending on the initial glass composition and temperature.

12.4.1 Mechanism of Crystallization

The stoichiometric composition of lithium disilicate ($Li_2Si_2O_5$) comprises 33.33 mol% LiO_2 and 66.67% mol SiO_2. The crystalline structure of lithium disilicate is related to phyllosilicate structures. Considerable improvements can be obtained using Al_2O_3 and K_2O additives to the stoichiometric glasses to be suitable as a restorative material in dental prostheses systems.

The lithium disilicates are known to crystallize by either surface crystallization or volume crystallization according to the starting glass chemical composition. Some components favor the crystallization by surface crystallization such as when B_2O_3 is incorporated.

The glass-ceramic pellets are made by pressing the starting amorphous glass powder being simultaneously sintered and crystallized. Additionally, the lithium disilicate composition can be modified to include La_2O_3, MgO and ZnO, B_2O_3, and Al_2O_3.

12.5 Chemical Composition of Lithium Disilicate Glass Ceramics

Whereas lithium disilicate is represented by the simple chemical formula Li_2O–$2SiO_2$, the actual chemical composition used in dental application contains a range of other oxides so as to achieve the desired properties. Glasses with compositions near the lithium disilicate stoichiometry (Li_2O–$2SiO_2$) need very high temperatures for melting and therefore tend to devitrify. Thus, there are difficulties in making a glass ceramic from these base glasses because they are exceptionally unstable. In addition, controlling the crystallization process of glass ceramics made from these compositions is difficult and the residual glassy phase is unstable.

To make these compositions behave consistently for production, it is necessary to control the developed phases and control the thermal behavior. In order to produce a finely divided crystalline lithium disilicate phase distributed in the glass matrix, the composition of the stoichiometric glass needs to be modified. One way this can be done is to add P_2O_5, which is frequently utilized as a nucleating agent in the manufacture of lithium disilicate glass ceramic. Although controlling the microstructure is possible by chemical composition modification, of course one has to realize that other properties may also be affected.

Silica, lithium oxide, alumina, potassium oxide, and phosphorus pentoxide, as well as an array of other useful components, can be used in the preparation of lithium disilicate-based glass ceramics. Li_2O and SiO_2 are the basic components in the crystallization of the required amount of the lithium disilicate phase. Additionally, BaO and Cs_2O can be used to stabilize the residual glass phase and improve its refractive index to match that of lithium disilicate and thus improve the translucency. Al_2O_3 and B_2O_3 of less than 3% are deemed to make chemically durable glass ceramics as these will lower the chemical solubility, which is an important factor for the application in the oral cavity. Calcium oxide is often added to compensate for the viscosity increase in the molten glass as a result of the addition of alumina.

Suitable additions of alkali oxides such as Na_2O, K_2O, and Cs_2O and alkaline earth oxides such as CaO and BaO lower the viscosity of the glass matrix and reduce the processing temperatures in order to achieve low fusion dental ceramics. A high

Table 12.1 Lithium disilicate glasses and glass ceramics after Peall (1993)

Glass composition in wt%			
Oxides	Glass 1	Glass 2	Glass 3
SiO_2	69.3	74.2	76.2
Li_2O	15.4	15.4	17.6
K_2O	6.05	3.25	2.23
ZnO	5.28		1.9
MgO			0.72
Al_2O_3		3.54	
P_2O_5	3.84	3.37	2.94
Melting temperature	1,450°C		
Annealing	450°C		
Heat treatment	480/1 h	650/4 h	650/4 h
	800/4 h	950/4 h	850/4 h
Glass ceramics			
Modulus of rapture PSI	28.5	31.4	29
TEC $\times 10^{-7}$/°C	127	78	98.5
Toughness, MPa $m^{0.5}$	2.8	3.30	2.79

strength lithium disilicate glass ceramic necessarily contains CaO to improve the flow. A small addition of calcium oxide improves the flow during glass casting and mostly affects the chemical durability together with sodium and barium oxides. The inclusion of K_2O does not show any significant change in the chemical solubility during the processing of lithium disilicate glasses. A combination of Y_2O_3, Ce_2O_3, and Eu_2O_3 are used to adjust the refractive index as well as to impart fluorescence, similar to the natural teeth. Nb_2O_5 and Ta_2O_5 are believed to modify the refractive index and improve the nucleation and chemical solubility of the glass ceramics.

In the case of castable lithium disilicate glass ceramics, the compositions essentially contain oxides of lithium, silicon and in addition, contain aluminum, calcium, and phosphorus, titanium, and/or zirconium oxides. Each of the elements in the compositions contributes to one or more desirable function. Lithium oxide acts as a powerful flux, which enhances the fluidity and castability of the glass during the processing. The proportion of lithium to silicon oxides is preferably chosen to approach as closely as possible the development of either lithium metasilicate or lithium disilicate.

Glass coloring oxides, coloring compounds, or their mixtures are added to the chemical composition of glass ceramic in order to obtain a glass ceramic color and fluorescence similar to that of the natural dental teeth. Certain oxides such as CeO_2, MnO_2, V_2O_5, Fe_2O_3, NiO_2, TiO_2, or combinations of two or more of the oxides are added during the melting of the parent glass compositions to serve only in making colored shades. It is important to consider carefully the usefulness of TiO_2 in that it acts mainly as a nucleating agent but, in combination with the other oxides, it can also act as a color component. In addition, other elements such as platinum and niobium oxide can be used to improve the flow and help to produce a very fine and uniform distribution of crystals. Some examples of lithium disilicate glasses and glass ceramics are presented in Table 12.1.

The refractive index of the lithium disilicate (~1.55) is close to that of the matrix glass (~1.5), a fact that allows for the possibility to make glass ceramics with a very high translucency. Specifically, the refractive index of the glass matrix can be increased to match that of the lithium disilicate phase by modifying the chemical composition slightly through doping with small amounts of heavy metal oxides such as Sr, Y, Nb, Cs, Ba, Ta, Ce, and/or Eu oxides, which will simultaneously act as coloring components.

Normally when lithium disilicate crystallizes it would show no tendency toward amorphous phase separation in the glass. This is changed with the addition of as small an amount as 2 mol% of P_2O_5, which leads to amorphous phase separation. The addition of P_2O_5 increased the nucleation rate of the stable lithium disilicate phase, and induces the amorphous phase separation and reduces the crystal growth rate. P_2O_5 combines with Li^+ cations to form a Li_3PO_4 phase glass, which that acts as nuclei for crystallization of lithium disilicate. P_2O_5 enhances the formation of metastable α and β lithium disilicate phases at the early and intermediate stages of heat treatment during the crystallization. Theses phases appear with an addition of P_2O_5 of >1 mol%. The use of P_2O_5 as the nucleating agent in lithium disilicate glass ceramics has the added benefit of raising the strength, which is something not seen with other nucleating agents and thus makes it the preferred nucleating agent.

The addition of oxides can produce a wide variety of effects on a wide range of properties of the lithium disilicate glass ceramic. It is down to the ingenuity of the researcher to manage the process of chemical composition modification and achieve the compromise of the functional properties.

12.6 The Properties of Lithium Disilicate Glass Ceramics

Lithium disilicate glass ceramic materials can be easily shaped by machining and subsequently improved into shaped products with high strength. The more energy is required to machine a ceramic the more likely it is that this results in uncontrolled fracture. Lithium metasilicate crystals have a lamellar or platelet form as shown in Fig. 12.2. It is this crystal structure that imparts a very good machinability and high edge strength of the lithium metasilicate glass ceramic.

The lithium metasilicate glass ceramic can be prepared by a process, which comprises (1) the production of a starting glass containing the components of the glass ceramic, (2) subjecting the starting glass to a nucleation heat treatment of 500–600°C determined by DTA to form nuclei suitable for forming lithium metasilicate crystals, and (3) subjecting the glass to a second heat treatment at temperature, which is higher than the first temperature to obtain glass ceramic with lithium disilicate as the main crystalline phase. Selecting the right composition and nucleation heat treatment will ensure very satisfactory crystal growth with a very homogeneous lithium disilicate structure. The XRD pattern of lithium disilicate is shown in Fig. 12.3.

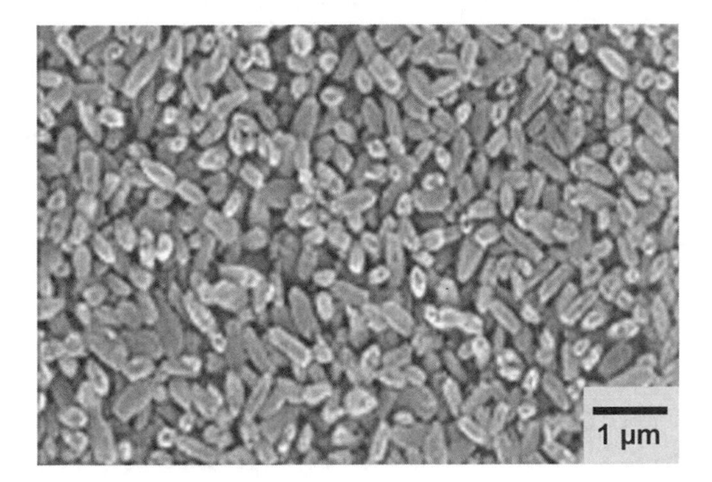

Fig. 12.2 Lithium metasilicate glass ceramics

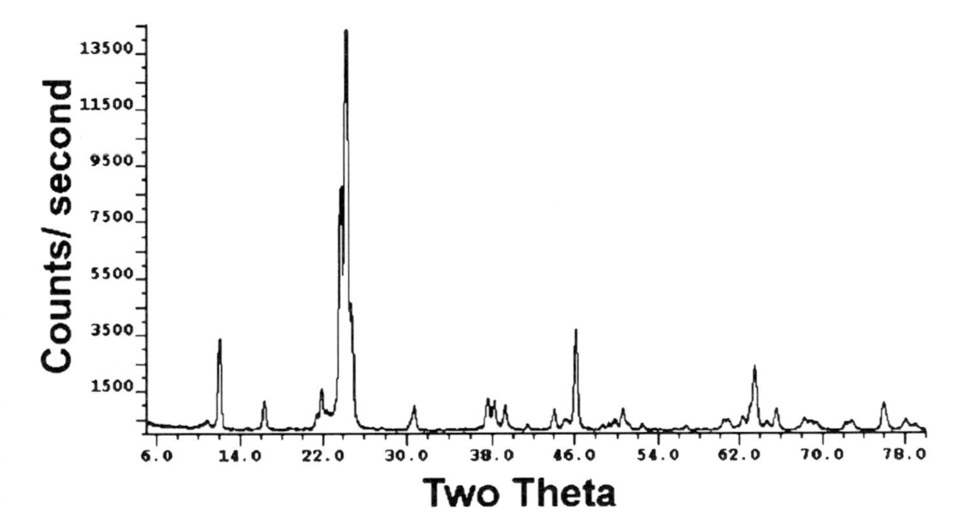

Fig. 12.3 The XRD pattern of lithium disilicate

The microstructure of the pressable lithium disilicate, shown in Fig. 12.4, consists of needle-like lithium disilicate crystals that are randomly distributed in the glassy matrix. The microstructure is free from defects and the size of crystals is approximately 3–6 μm in length. The high aspect ratio of the lithium disilicate crystals is believed to enhance the mechanical properties, i.e., strength and fracture toughness of the glass-ceramic.

Lithium disilicate glass ceramics have important properties, as a result of which they are particularly suitable for use as a dental materials. The lithium disilicate glass-ceramics not only have very high strength in the range of 200–400 MPa

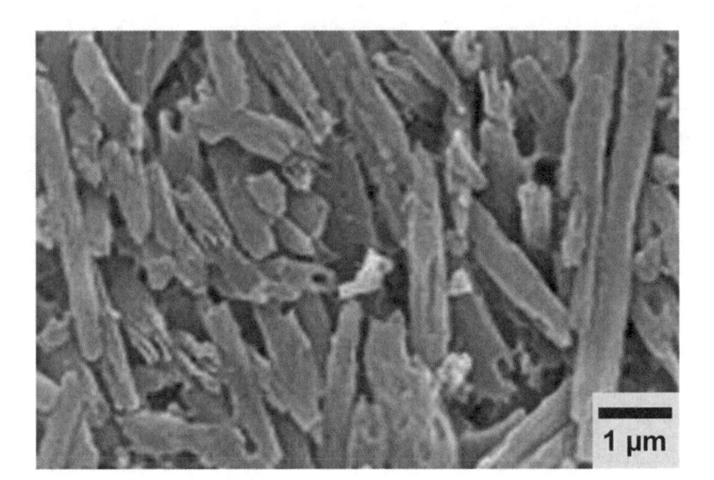

Fig. 12.4 Lithium disilicate glass ceramics

Table 12.2 Properties and chemical composition of lithium disilicate

Properties		Chemical composition	
CTE × 10^{-6}/K (100–500°C)	10.5	Oxide	wt%
Biaxial flexural strength, MPa	400	SiO_2	68.7
Fracture toughness, MPa $m^{0.5}$	2.75	Na_2O	1.5
Modulus of elasticity, GPa	95	Al_2O_3	4.8
Vickers hardness, MPa	5,800	CaO	1
Chemical resistance, μg/cm^2	40	BaO	2.8
Maturing temperature	920	Li_2O	14.4
		K_2O	2.2
		P_2O_5	3.3

but also a high fracture toughness value in the range of 2.5–4.5 MPa $m^{0.5}$. Lithium disilicate, with a suitable addition of other metal oxides, also has a translucency similar to that of the natural tooth. The composition and properties of an example lithium disilicate glass ceramic are shown in Table 12.2. The high mechanical strength of lithium disilicate qualifies the material to make strong products for use as anterior and posterior crowns.

12.7 Problems Encountered with Lithium Disilicate

As is always the case, lithium disilicate glass ceramics do have some weaknesses. Many compositions of the lithium disilicate glass ceramics lack the necessary chemical resistance for use as dental material, when it is permanently being flushed with fluids of various kinds in the oral cavity.

Conventional lithium disilicate glass ceramics do not satisfy various requirements for the processing of dental products by plastic deformation. Any attempt at producing dental restorations of the desired shape using the casting route tends to give rise to a range of problems. The viscosity is not easily adjusted, particularly when processed in the plastic state to form a shaped dental product. This means that lithium disilicate may not easily flow and unfortunately it also reacts with investment materials so casting is a problem. The requirement that the glass ceramic should flow in the plastic state in a controlled manner and at the same time react only to a small extent with the investment material can be achieved by the use of La_2O_3 and/or Al_2O_3 in the appropriate quantities. Such a glass ceramic is free flowing and can be pressed in the plastic state to yield a glass ceramic material.

Furthermore, lithium disilicate has poor dimensional stability on heating resulting in the deformation of the dental restoration during processing. The glass, once it is cast, must subsequently be heat treated to crystallize the lithium disilicate phase. This may result in microstructural inhomogeneity which may produce something that is not expected by the manufacturer.

Alternatively, glass can be crystallized in bulk and subsequently milled into powder, but in this case the maturation temperature will be shifted higher and it will be difficult to achieve homogeneity when adding pigments and other additives. There will also be a difficulty in producing bodies with reliable translucency by either partial or full sintering.

The presence of ZnO in lithium silicate glass ceramics is undesirable especially when highly translucent dental restorations are to be produced. Under such circumstances, the strong opalescent effect caused by ZnO is apparent and results in unacceptable optical properties for a restoration, which is to imitate natural tooth material.

Lithium disilicate is hard on machining tools during CAD–CAM processing as a result of its high strength and toughness and the machined restorations may have poor edge strength. There exist compositions, which crystallize directly in the form of lithium disilicate from the precursor glasses. Other compositions can crystallize first into metastable lithium metasilicate phase that is then converted to crystalline lithium disilicate glass ceramic with very high strength. The glass ceramic, which contains lithium metasilicate as the main phase has a reduced strength and is more easily machinable compared to glass ceramics that only contains the lithium disilicate phase. The lithium metasilicate crystal structure, Li_2SiO_3, is strong enough to be easily milled without excessive wear. The microstructure shown in Fig. 12.2 contains 40 vol% lithium metasilicate crystals with an approximate crystal size of 0.5 μm.

The best way to solve the problem of machinability of lithium disilicate is to use a starting glass of composition that crystallize first into the metastable lithium metasilicate (Li_2SiO_3) glass ceramic as main crystalline phase rather than lithium disilicate. This lithium metasilicate glass ceramic can be easily machined into the desired shape restorations, and then converted by a heat treatment into a lithium disilicate glass ceramic after a very limited shrinkage with outstanding mechanical properties, excellent optical properties, reduced opalescence, and very good chemical stability.

The commercially available IPS e.max CAD (Ivoclar-Vivadent, Liechtenstein) uses this approach very effectively. Lithium metasilicate glass ceramic blocks are processed into dental restorations by the CAD/CAM technique. At this stage the lithium metasilicate glass ceramic has a strength of about 80–150 MPa. The dental restoration is then subjected to a further heat treatment between 700 and 1,000°C to obtain the desired lithium disilicate microstructure, which has a much higher strength of 200–400 MPa and will also be tooth-like in color. The product can be further veneered with a glass or a glass ceramic by sintering or hot pressing technique to obtain the final dental restoration with natural esthetics.

Further Reading

Borom, M.P., Turkalo, A.M.: Strength and microstructure in lithium disilicate glass-ceramics. J. Am. Ceram. Soc. **58**, 9–10 (1975)

Brodkin, D., Panzera C., Panzera P.: Pressable lithium disilicate glass ceramics. US patent 6,455,451, 2002

Doherty, P.E., Lee, D.W., Avis, R.S.: Direct observation of the crystallization of Li_2O-Al_2O_3-SiO_2 glasses containing TiO_2. J. Am. Ceram. Soc. **50**(2), 77–81 (1967)

Hench, L.L., Freiman, S.W., Kinser, D.L.: Phys Chem. Glasses **12**, 58 (1971)

Höland, W., Schweiger, M., Frank, M., et al.: A comparison of the microstructure and properties of the IPS Empress 2 and the IPS Empress glass-ceramics. J Biomed. Mater Res. **53**, 297–303 (2000)

Iqbal, Y., Lee, W.E., Holland, D., James, P.F.: Crystal nucleation in P2O5-doped lithium disilicate glasses. J Mater Sci **34**, 4399–4411 (1999)

Kim, S.S., Sanders, T.H.: Thermodynamic modeling of phase diagrams in binary alkali silicate systems. J. Am. Ceram. Soc. **74**(8), 1833–1840 (1991)

Kracek, F.C.: The cristobalite liquidus in the alkali oxide-silica systems and the heat of fusion of cristobalite. J. Am. Chem. Soc. **52**(4), 1436–1442 (1930)

Moriya, Y., Warrington, D.H., Douglas, R.W.: Phys Chem. Glasses **8**(1), 19–25 (1967)

Peall, G.H.: Lithium disilicate containing glass ceramics some of which are self glazing. EP 536 479 A1, 1993

Pierson, J.E.: Lithium disilicate containing glass ceramics some of which are self glazing. EP 0536479 A1, 1992

Samsonov, G.V.: The Oxide Handbook, 2nd edn. IFI/Plenum, New York (1982)

Schweiger, M., Frank, M., Rheinberger, V., Höland, W.: Sinterable lithium disilicate glass ceramic. US patent 5,968,856, 1999

Schweiger, M., Frank, M., Rheinberger, V., Höland W.: Lithium disilicate glass ceramics dental product. US patent 6,342,458, 2002

Schweiger, M., Frank, M., Rheinberger, V., Höland, W.: Sinterable lithium disilicate glass ceramic. US patent 6,514,893, 2003

Part V
Bioactive Glass and Bioactive Glass Ceramics

When an inert ceramic is implanted to restore damaged bone tissues in the human body, the implant is generally surrounded by membranes of collagen fibers and isolated from the bones. When bioactive glasses and bioactive glass ceramics are implanted in the human body, no isolation exists by such fibrous membranes as the bioactive glass can bond to bones strongly and naturally. For this reason, bioactive glasses and bioactive glass ceramics with specified phases represent suitable materials for restoring bone damage.

Chapter 13
Bioactive Glasses

13.1 Nature of Bioactive Glass

Bioactive glasses were discovered by Hench in 1969 and provided for the first time an alternative interfacial bonding of an implant with host tissues. Bioactive glasses have osteoinductive and osteoconductive properties that can be used for the repair and reconstruction of damaged bone tissues. They are surface-reactive materials designed to induce biological activity and form a strong bond with the living tissues such as bone.

Bioactive glasses, based on silicate compositions, have therefore found medical applications as synthetic bone grafts for general orthopedic, craniofacial, maxillofacial, and tissue engineering scaffolds for bone. Silicate bioactive glasses efficiently aid osteogenesis in the physiological system via the exposure of the primary body osteoblasts to the ions dissolved from the surface of bioglasses.

Bioactive glasses have been known for many years by their ability to react with body tissues and form a biologically active surface layer of hydroxycarbonate apatite (HCA) on exposure to physiological fluids. The HCA layer which is responsible for interfacial bonding are reinforced by collagen fibers and are chemically and structurally equivalent to natural bone.

The bioactive glass structure, like any glass structure, consists of network forming oxide (SiO_2) and some highly selective network modifier oxides, including Na_2O, K_2O, CaO, MgO, and includes two types of bonds namely bridging oxygen bonds (Si–O–Si) and nonbridging oxygen bonds (Si–O–NBO).

The bridging oxygen bonds occur between two neighboring silicon atoms and holds the network together; while the nonbridging oxygen bonds occur between Si and a network modifying atom (Na^+, K^+, Ca^{2+}, etc). This fact helps to disrupt the continuity of the glass network (Si–O–Si bonds) through the surface exchange of silicon by alkali cations leading to the formation of nonbridging oxygen bonds Si–O–NBO. The relative proportion of bridging oxygen bonds to nonbridging oxygen bonds is expected to control the surface reactivity of bioglass in the physiological fluids.

E. El-Meliegy and R. van Noort, *Glasses and Glass Ceramics for Medical Applications*,
DOI 10.1007/978-1-4614-1228-1_13,

The network former (SiO_2) in the bioactive glass holds the three-dimensional nonperiodic glass structure together during selective dissolution of cations (e.g., Na^+) by suppressing the detachment of some other ions. The presence of SiO_2 also helps the precipitation or surface reconstruction of the loose silica-rich layer, and hence enhances the formation of the hydroxyapatite layer.

Bioactive glasses attach to human tissues through the formation of hydrated silica with high surface area upon which a polycrystalline HCA biolayer is formed. The HCA is chemically and structurally equivalent to the mineral phase in bone and is responsible for interfacial bonding. A small change in the chemical composition can produce a drastic effect on the biological behavior, whether it is inert, resorbable, or bioactive indicating the importance of minor changes in the chemical components.

13.2 Chemical Composition of Bioactive Glasses

Hench used the Na_2O–CaO–SiO_2 diagram in "Phase Diagrams for Ceramics" to design the first three bioactive glass compositions (Fig. 13.1). The composition of 45% SiO_2–24.5% Na_2O–24.5% CaO–6% P_2O_5 was selected to provide a large amount of CaO with some P_2O_5 in a Na_2O–SiO_2 matrix. The composition is very close to a ternary eutectic temperature and thus is easy to melt. A bioactive glass containing a higher content of Na_2O developed by Hench with a composition of 45 wt% SiO_2, 24.5 CaO, 6 P_2O_5, and 24.5 Na_2O, is good for making the glass at a lower melting temperature.

The bioactive glass is found to be strongly bonded to tissues when implanted in a rat femur for 6 weeks. However, the high concentration of Na_2O (10–35mol%) in

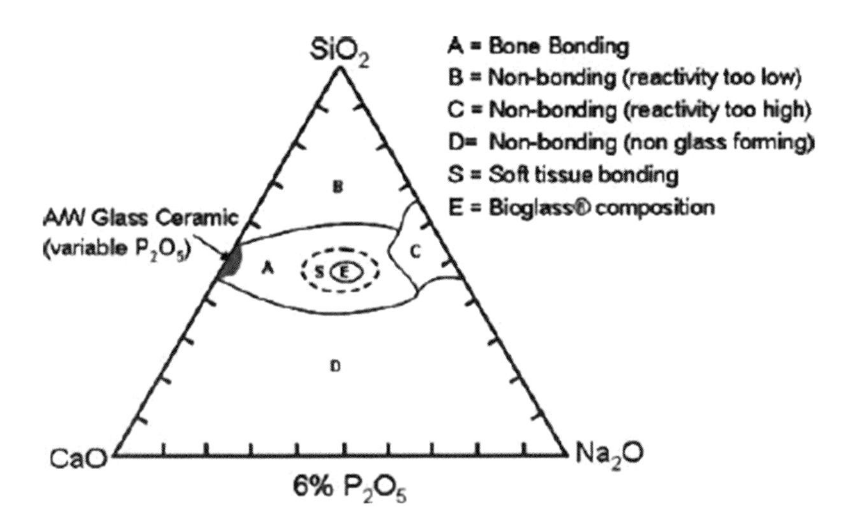

Fig. 13.1 Compositional diagram of bioactive glasses for bone bonding

most commercial bioactive glasses may reduce the bioactivity in vivo and diminishes the difference between the glass transition temperature (T_g) and the onset of crystallization leading to a small working temperature range especially during coating applications. The glass transition temperature, the softening temperature, and the crystallization temperature are 500, 560, and 710°C respectively.

The in vitro tests showed that the 45S5 Bioglass™ composition developed a hydroxyapatite layer in test solutions that did not contain calcium or phosphate ions.

Silicate bioactive glasses mostly contain SiO_2, Na_2O, CaO, and P_2O_5, besides some necessary oxides to induce certain required properties such as MgO, K_2O, or B_2O_3. The rapid bioactive behavior of silicate bioactive glasses is related to the role of silicon in the surface reactions in vitro and in vivo. A lower content of SiO_2 results in a lower network connectivity, which is crucial for the bioactive glass surface reactivity and ionic exchange on the surface. In addition, the presence of MgO promotes the formation of apatite and its solubility in vitro.

To improve the suitability of bioactive glasses for in vivo applications with good levels of bioactivity, it is therefore desirable to seek bioactive glasses with lower levels of Na_2O and little or no K_2O that can be formulated for a wide range of applications. Bioactive glasses with content of Na_2O lower than 10 mol% and free of K_2O and Al_2O_3 based on the system SiO_2–Na_2O–CaO–MgO–P_2O_5 provide a suitable rate of apatite deposition for potential bioactivity and rapid reconstruction of damaged bone.

Compositions with greater than 60 wt% SiO_2 are expected to be bioinert and those with between 52 and 60 wt% SiO_2 exhibit slower bonding rates. The bioactive glasses comprise SiO_2 (40–52%), CaO (10–50%), Na_2O (10–35%), P_2O_5 (2–8%), CaF_2 (0–25%), and B_2O_3 (0–10%). A small change in the composition produces a drastic effect on the behavior of bioglasses. The bioactive glasses are distinguished by less than 60 mol% SiO_2, high Na_2O and CaO content, and high molar ratio of CaO/P_2O_5.

Sodium ions in the bioactive glass composition can also contribute as an effective flux, lowering the glass melting temperature, and increase dissolution from the bioactive glass surface, resulting in the formation of a silica-rich layer upon which successive calcium phosphate layer precipitate that finally crystallize into HCA, necessary for bonding of the implant to tissues.

In addition, leaching of Na^+ also affects the physiological balance at the bioactive glass interface and increases temporarily the local pH that promote the synthesis and cross-linking of collagen and the formation of hydroxyapatite as a necessary effect for in vivo bone growth and repair.

MgO contents in the SiO_2–CaO–MgO–P_2O_5 system promote in vitro apatite formation and leads to the formation of a Mg-rich calcium phosphate layer. Magnesium affects the crystallinity and solubility of apatites and help to regulate the dissolution of the apatite precipitates on the bioactive glass.

The disadvantage of bioactive glass that limits its application is its brittleness such that limits its application is its handling and mechanical properties are not adequate for significant load bearing applications. This fact makes the use of bioactive glass ceramics, with specific crystalline phases, more attractive in many applications as discussed in Chap. 14.

13.3 Properties of Bioactive Glasses

The essential advantage of a bioactive glass is that it can be used as a reference for the development of new bioactive materials, e.g., a bioactive glass-ceramic with a specified crystallinity or a composite with specified mechanical strength based on that glass. The rapid bioactive behavior in silicate-based glasses has been related to the role of SiO_2 or silicon in their surface reactions and therefore on their in vivo and in vitro behavior.

A particular advantage of a bioactive glass is the ability to bond to both hard and soft tissues. The surface reactions that occur in the bioactive glasses allow the subsequent crystallization of apatite crystals, cell adhesion, and collagen formation. Another advantage of silicate-based glasses is that they show a genetic control of the cellular response of osteoblasts. The different genes are regulated within a short time of exposure of primary human osteoblasts to the ionic dissolution products of bioactive glasses. Biologically active glasses have osteoinductive and osteoconductive properties that are ideally suited to the repair and reconstruction of damaged bone. Osteoinductive materials stimulate the osteoprogenitor cells to differentiate into osteoblasts that then begin new bone formation, while osteoconductive materials serve as a substrate upon which osteoblasts are able to spread and generate new bone.

Bioactive glasses have been known for many years to be capable of reacting biochemically and continue to form a surface layer of biologically active HCA on exposure to physiological fluids. The HCA layer reinforced by collagen fibers, responsible for interfacial bonding, is chemically and structurally equivalent to the natural bone. The HCA phase that forms on bioactive implants is equivalent chemically and structurally to the mineral phase in bone. It is that equivalence which is responsible for interfacial bonding.

In many cases, the interfacial strength of adhesion is equivalent to or greater than the cohesive strength of the implant or the tissue bonded to the bioactive implant with adherence at the interface sufficient to resist mechanical fracture. In porous constructs where good biological fixation relies on tissue ingrowth, the requirement is that pores are greater than 150 μm in diameter to provide a blood supply to the host tissues.

Bonding of bioglasses with the natural bone results from a series of physicochemical reactions. On exposure to physiological fluids, the interstitial cation exchange occurs through the replacement of Na^+ and Ca^{2+} in the glass surface by protons (H^+ or H_3O^+) forming surface silanol groups and nonstoichiometric hydrogen-bonded compounds. The interfacial pH becomes more alkaline and the concentration of surface silanol groups increases, resulting in the condensation of silanol species into a silica-rich layer. The silica layer has a microporous structure with high specific surface area and acts as a buffer layer between the bioactive glass and the newly formed apatite in vitro and lies within the gel layer in vivo.

When the bioactive glass is implanted in vivo, selective diffusion by the partial dissolution of the bioactive glass lattice is the principal step in the formation of the superficial silica layer upon which the bone mineralization forms. The calcium and phosphorous ions recrystallize into HCA. As the crystalline layer forms, the body's proteins are attracted and bind to the HCA layer, which is thought to be the mechanism for bone growth.

13.4 Bioactivity of Bioactive Glasses

A minimum rate of apatite formation is necessary for the glass to be bioactive and bond with living tissues. The rate of HCA formation on the glass surface is a strong indication as to the levels of in vitro bioactivity, which can be examined by the immersion of bioactive glasses in simulated body fluid (SBF) for different time-periods.

A variation in pH and weight loss during immersion of bioactive glasses in SBF solution in vitro is due to the conversion reactions. The conversion reactions involve the surface ionic exchange between the glass and SBF followed by the precipitation and crystallization of HCA. The exchange modifies the network by forming non-bridging weak oxygen bonds and increases the interfacial pH.

The chemical composition of an example of a bioactive glass based on the system SiO_2–Na_2O–CaO–P_2O_5–MgO with a very good bioactivity is shown in Table 13.1. The DTA in Fig. 13.2 shows high transition and softening temperatures and in turn a high crystallization temperature. The surface of the bioactive glass would appear to have excellent solubility during immersion in SBF, indicating high reactivity. The solubility has been calculated to be 3.32×10^{-5} g/cm^2. In addition, a strong dissolution of Si^{4+}, Mg^{2+}, Ca^{2+}, and Na^+ from the bioactive glass was seen to take place with its immersion in SBF.

Table 13.1 Chemical composition of bioactive glasses

	wt%				
Oxides	P_2O_5	SiO_2	CaO	Na_2O	MgO
Modified	10.3	50.1	24.2	10.3	5.2

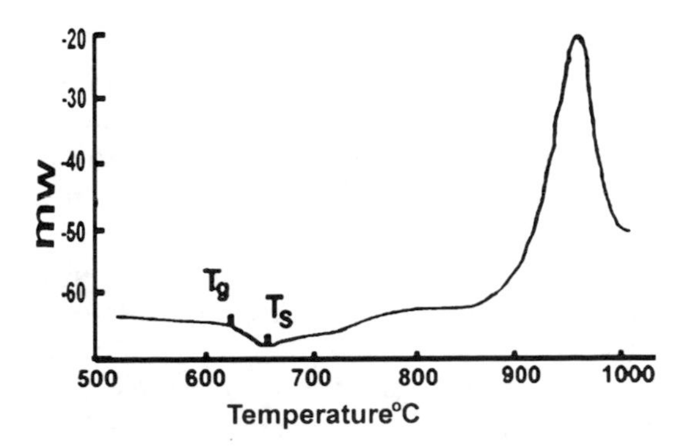

Fig. 13.2 DTA of bioactive glass

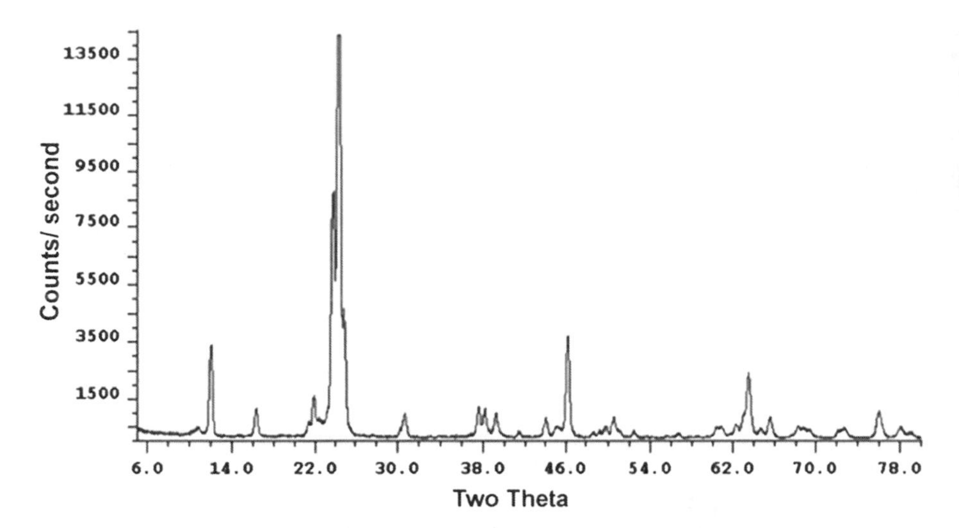

Fig. 13.3 Bioactive glass after 4 weeks immersion in SBF

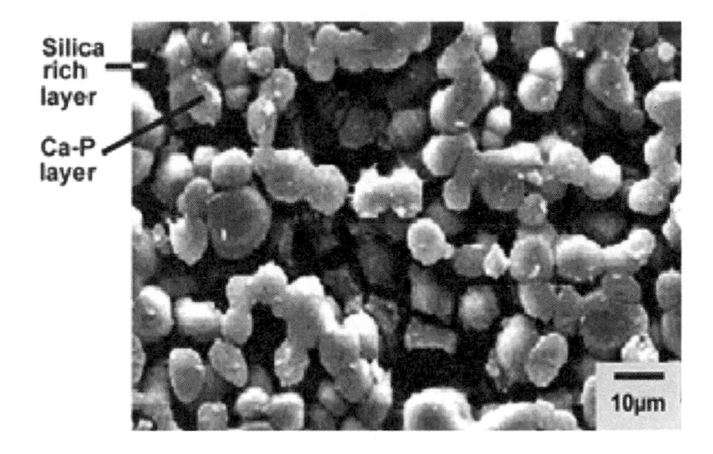

Fig. 13.4 Bioactive glass after 12 weeks implantation in rat's femur (in vivo)

Of course the bioactive glass has a high fusion temperature and relatively high transition and softening temperatures. There is a possibility for this bioactive glass to be modified to reduce the fusion temperature without affecting the bioactivity. Other chemical components can be added to perform the respected modification such as MgO or B_2O_3, CaF_2 or MgF_2.

Bioactive glasses with a lower content of Na_2O and free of K_2O and Al_2O_3 are expected to provide a high rate of apatite deposition. In addition, these bioactive glasses may help rapid reconstruction of damaged bone and suit different coating applications. The values of the transition temperature, softening temperature, and the crystallization temperature are 620, 645, and 945°C, respectively as shown by DTA in Fig. 13.2.

Postimplantation of a particulate bioactive glass the in rat femur in vivo, revealed normal healing of the soft tissues with no abnormalities over the site of the operation as shown in Fig. 13.3. The critical size bone defects created in the rat femur and grafted with modified bioglass particles at 12 weeks postimplantation, were completely healed by filling with mineralized bone matrix as shown in Fig. 13.4. The elemental analysis by EDX showed that the Ca/P molar ratio (~1.67) had nearly the same value as for hydroxyapatite ($Ca_{10}(PO_4)6(OH)_2$) of the natural bone.

Further Reading

Cerruti, M., Greenspan, D., Powers, K.: Effect of pH and ionic strength on the reactivity of Bioglass 45 S5. Biomaterials **26**, 1665–1674 (2005)

El-Meliegy, E.M., El-Bassyouni, G.T.: Study of the bioactivity of fluorophlogopite-whitlockite ceramics. Ceram Int **34**(6), 1527–1532 (2008)

EL-Rashedi, A.M.: Preparation and characterization study of bioactive glasses for orthopaedic applications, PhD Thesis, Al-Azhar University, Cairo, Egypt

Franks, K., Abrahams, I., Knowles, J.C.: Development of soluble glasses for biomedical use Part I: In vitro solubility measurement. J Mater Sci Mater Med. **11**(10), 609–614 (2000)

Gross, U.M., Strunz, V.: J Biomed Mater Res **14**, 607 (1980a)

Gross, U.M., Strunz, V.: The anchoring of glass ceramics of different solubility in femur of the rat. J. Biomed. Mater. Res. **14**, 607 (1980b)

Hench, L.L.: Bioceramics: from concept to clinic. J Am Ceram Soc **74**, 1487 (1991)

Hench, L.L.: The story of Bioglass. J Mater Sci Mater. Med. **17**, 967–978 (2006a)

Hench, L.L.: The story of bioglass. J Mater Sci: Mater Med **17**, 967–978 (2006b)

Hill, R.G., Stevens, M.M.: Bioactive glass. US patent 0208428, 2009

Holand, W.: Biocompatible and bioactive glass-ceramics – state of the art and new directions. J Non Cryst Solids **219**, 192–197 (1997)

Knowles, J.C., Franks, K., Abrahams, I.: Investigation of the solubility and ion release in the glass system K_2O-Na_2O-CaO-P_2O_5. Biomaterials **22**, 3091–3096 (2001)

Kokubo, T.: In: Hench, L.L., Wilson, J. (eds.) An Introduction to Bioceramics, p. 75. World Scientific, Singapore (1993)

Kokubo, T., Takadama, H.: How useful is SBF in predicting in-vivo bone bioactivity. Biomaterials **27**, 2907–2915 (2006)

Marikani, A., Maheswaran, A., Premanathan, M., Amalraj, L.: Synthesis and characterization of calcium phosphate based bioactive quaternary P2O5 – CaO – Na2O–K2O glasses. J. Non Cryst. Solids **354**, 3929–3934 (2007)

Nakamura, T., Yamamuro, T., Higashi, S., Kokubo, T., Itoo, S., Biomed, J.: Mater Res. **19**, 685 (1985)

Piotrowski, G., Hench, L.L., Allen, W.C., Miller, G.J.: Mechanical studies of the bone bioglass interfacial bond. J Biomed Mater Res Symp. **9**, 47 (1975)

Wilson, J., Yli-Urpo, A., Risto-Pekka, H.: In: Hench, L.L., Wilson, J. (eds.) An Introduction to Bioceramics, p. 63. World Scientific, Singapore (1993)

Yamamuro, T.: A/W glass-ceramic: clinical applications. In: Yamamuro, T., Hench, L.L., Wilson, J. (eds.) Handbook of Bioactive Ceramics. Bioactive Glass and Glass Ceramics, vol. 1, p. 89. CRC, Boca Raton, FL (1990)

Yamamuro, T.: In: Hench, L.L., Wilson, J. (eds.) An Introduction to Bioceramics, p. 89. World Scientific, Singapore (1993)

Yoshii, S., Kakutani, Y., Yamamuro, T., Nakamura, T., Kitsugi, T., Oka, M., Kokubo, T., Takagi, M.: J Biomed Mater Res **22A**, 327 (1988)

Chapter 14
Models of Bioactive Glass Ceramics

It was shown very early on in the development of bioactive glasses that when a bioactive glass, with a composition suitable for a glass ceramic, is heat treated to create the crystalline phase, the material can retain its bioactivity. This was an important finding as bioactive glasses are brittle and lack strength and toughness. As a consequence, the handling and mechanical properties of bioactive glasses are not adequate for significant load bearing applications and the monophase bioactive glasses are generally restricted to clinical situations, which are nonload bearing.

Many examples of bioactive glass ceramics based on silicate compositions have since been developed, including apatite glass ceramics, apatite wollastonite glass ceramics, apatite fluorophlogopite glass ceramics, apatite mullite glass ceramics, fluorcanasite glass ceramics, and fluorrichterite glass ceramics. These bioactive glass ceramics are being considered for a wide range of applications, including middle ear implants, alveolar ridge augmentation, scaffolds for tissue engineering, and the treatment of dentin sensitivity.

14.1 Apatite Glass Ceramics

In 1977 a bioactive apatite glass-ceramic, based on the 45S5 bioactive glass formula of Hench but with small additions of K_2O and MgO, was first implanted in animal models by Professor Ulrich Gross and colleagues. They found that the glass-ceramic bonded to bone with a mechanically strong interface and was so bioactive that within 1 h of implantation a hydroxycarbonate apatite (HCA) layer, some 500 nm deep was formed. Specimens soaked in simulated body fluid showed that it formed a layer of carbonate-containing hydroxyapatite of small crystallites and/or a defective structure on its surface in the fluid. However, additions of multivalent cations such as Ti and Ta to the glass composition prevented bonding.

E. El-Meliegy and R. van Noort, *Glasses and Glass Ceramics for Medical Applications*, DOI 10.1007/978-1-4614-1228-1_14,

Table 14.1 Chemical composition of two bioactive glasses in weight percent

Glass	SiO_2	CaO	MgO	CaF_2	Na_2O	K_2O	P_2O_5	Ta_2O_5	La_2O_3
AP40	44.00	32.16	2.84	5.00	4.60	0.20	11.20	–	–
RKKP	43.94	31.93	2.79	4.94	4.55	0.19	0.27	0.99	0.50

This bioactive glass ceramic has since been marketed as Ceravital®. One of its first uses was as an implant for the reconstruction of the ossicular chain or the posterior canal wall in the middle ear. However, clinical use of this bioactive material was limited due to instability of the crystal phase boundaries in the glass-ceramic and late hearing loss resulted due to resorption of the prosthesis.

A fully crystallized bioactive glass-ceramic based on the system P_2O_5–Na_2O–CaO–SiO_2 named Biosilicate™ has been proposed to treat dentin hypersensitivity (DH) by HCA deposition in open dentinal tubules (Zanotto et al. 2004). The similarity of composition between bone, dentin, and enamel led to the assumption that bioactive glasses and glass-ceramics could be efficient for the regeneration of enamel and dentin. The Biosilicate™ has been shown to deposit a homogeneous layer of HCA in open dentinal tubules and the amount of apatite glass ceramic particles played an important role in reducing DH.

An interesting and valuable avenue of exploration is the idea of developing bioactive glass ceramics tailored for patients with particular problems such as osteopenia. The idea is that a subtle difference in the chemical composition of the glass ceramic may invoke a different biological response. Osteopenia refers to the early signs of bone loss that can develop into osteoporosis. With osteopenia bone mineral density is lower than normal. However, it is not yet low enough to be considered osteoporosis. A bioactive glass ceramic containing small additions of Ta_2O_5 and La_2O_3 was developed by Fini (1997) for the use in patients with osteopenia (Table 14.1). It has been shown that this modified glass ceramic, known as RKKP, performs well in osteopenic bone. Various studies have been carried out comparing RKKP with a glass ceramic without the additions of Ta_2O_5 and La_2O_3 (AP40) but so far no clear answer has been derived to explain this difference in behavior.

14.2 Apatite–Wollastonite Glass-Ceramics

One of the most important modifications of bioactive glasses was the development of A/W (apatite/wollastonite) bioactive glass-ceramics by Professors Yamamuro and Kokubo and colleagues at Kyoto University, Kyoto, Japan. A unique processing method produced a very fine-grained glass ceramic composed of very small apatite (A) and wollastonite ($W = CaSiO_3$) crystals bonded by a bioactive glass interface. Mechanical strength, toughness, and stability of AW glass-ceramics (AW-GC) in physiological environments are excellent. Bone was found to bond to A/W-GC implants with a high interfacial bond strength.

An apatite–wollastonite glass ceramic can be produced in the SiO_2–CaO–MgO–P_2O_5–F system. The chemical composition in wt% comprises 34.0SiO_2, 44.7CaO, 4.6MgO, 16.2P_2O_5, and 0.5CaF_2. One problem with A/W glass ceramics is that wollastonite tends to surface nucleate, such that it cannot be cast into a glass block and then crystallized. However, it can be made into a glass-ceramic by putting the glass powder through a firing cycle that combines densification with crystallization. The glass powder can be fully densified at approximately 830°C and oxyfluoroapatite, $Ca_{10}(PO_4)6(O, F_2)$ and wollastonite ($CaSiO_3$) precipitated with a heat treatment at approximately 870–900°C. The final glass ceramic can be produced such that it is free of cracks and pores. Crystals are homogeneously distributed in the glassy matrix with a size of approximately 50–100 nm. It is possible to crystallize approximately 38 wt% apatite and 24 wt% wollastonite according to X-ray diffraction.

In the reaction mechanism that characterizes the bioactivity of these glass ceramics, Kokubo and Takadama (2006) discovered that a dissolution process of Ca^{2+} ions takes place in the surrounding body fluid during immersion in simulated body fluid (SBF), which is attributed to the high CaO content of the glass ceramic. Hence, as the concentration of Ca^{2+} in the SBF rises, Si–OH groups on the surface of the glass-ceramic act as favorable sites for apatite nucleation. The Si–OH groups immediately combine with Ca^{2+} ions in the fluid to form an amorphous calcium silicate on the surface. After a long period, this calcium silicate combines with phosphate ions to form an amorphous calcium phosphate layer with a low Ca:P atomic ratio. This phase later transforms into bone-like apatite crystals, increasing its Ca:P ratio and incorporating minor ions such as Na^+, Mg^{2+}, and Cl^-. After the formation of apatite on the surface of the material, additional requirements of Ca^{2+} ions and phosphate groups are covered by mass transport from the surrounding body fluid. In these glass ceramics, however, Kokubo and Takadama (2006) did not observe the formation of a SiO_2 gel layer as is the case with the bioactive glasses from the SiO_2–CaO–Na_2O–P_2O_5 system.

A fine crystalline microstructure of the glass-ceramic is achievable, which is reflected in a bending strength of some 215 MPa, compressive strength of 1,080 MPa, and a fracture toughness of 2.0 MPa $m^{0.5}$. These good mechanical properties makes this glass ceramic potentially suitable for such applications as load bearing implants. The A–W glass ceramic is also machinable but its major drawback is the relatively low fracture toughness compared with other ceramics such as alumina or zirconia.

Nevertheless, A–W has been used as bone spacers and fillers in bulk and granular forms with dense and porous structures and also as artificial vertebrae, intervertebral disks and iliac crests in dense bulk form. Unfortunately, despite having a superior mechanical strength to many other biomaterials such as bioactive glasses, bioceramics, and bone, A–W glass ceramic cannot be used to repair bony defect in high stress bearing areas such as the femur or tibial bones, as its fracture toughness is too low and its elastic modulus too high.

Cerabone™ is a commercially available version of a high strength, bioactive glass-ceramic composed of apatite/wollastonite.

Table 14.2 Chemical composition of mica phosphate glass

Oxides	SiO_2	Al_2O_3	MgO	K_2O	MgF_2	Na_2O	CaO	B_2O_3	P_2O_5
F2	30.01	6.71	3.59	4.16	5.75	0.61	19.68	5.75	23.74

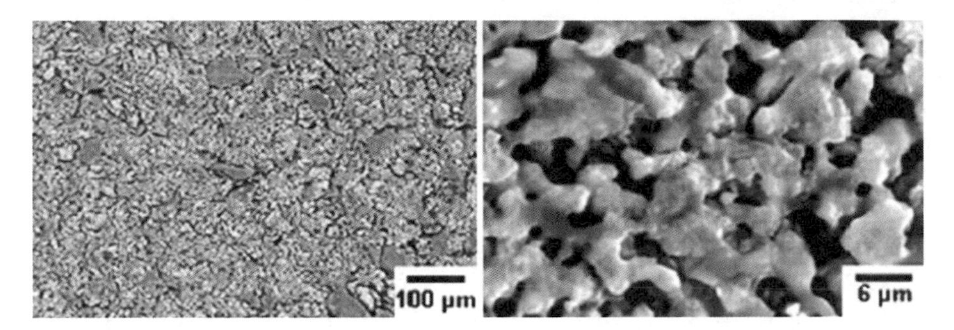

Fig. 14.1 SEM of fluorophlogopite–apatite soaked for 1 week in SBF

14.3 Apatite–Fluorophlogopite Glass-Ceramics

Fluorophlogopite mica can be made to crystallize simultaneously with apatite in the system SiO_2–Al_2O_3–MgO–CaO–Na_2O–K_2O–P_2O_5–F system by controlled crystallization and produces a machinable bioactive glass ceramic. These glass ceramic materials contain a minimum Al_2O_3 content as in the composition as shown in Table 14.2. The distinct crystal phases, that were precipitated simultaneously by bulk nucleation, included apatite an fluorophlogopite. Simultaneously with a bulk volume nucleation mechanism. The formation of apatite was achieved by influencing the miscibility of the glass to form a CaO-P_2O-rich glassy phase as the nucleus as shown in Fig. 14.1. The resulting glass-ceramic demonstrated bending strength of 120–180 MPa and fracture toughness of 1.2–2.5 MPa $m^{0.5}$.

Machinable bioactive glass-ceramics containing mica and apatite phases display good mechanical strength, fracture toughness, biocompatibility, and a satisfactory degree of bioactivity although they take a longer time to bond to living bone tissue than bioactive glasses and can be considered as class B bioactive implants (see Chap. 13). Good machinability is related to the uniform fine microstructure of the mica phase randomly distributed in the glassy matrix. This means that these materials can be tailored easily to produce surgical parts with various complex shapes by using CAD–CAM techniques.

14.4 Apatite–Mullite Glass-Ceramics

Glasses in the system $SiO_2 \cdot Al_2O_3 \cdot P_2O_5 \cdot CaO \cdot CaF_2$ will crystallize predominantly to apatite and mullite upon heat-treatment. Mullite or porcelainite is a rare silicate mineral and can form two stoichiometric forms, $3Al_2O_32SiO_2$ or $2Al_2O_3SiO_2$. Such ceramics are potentially bioactive and osseoconductive, and have a high resistance to fracture. The glass precursor exhibits a low liquidus temperature, is castable, and is readily crystallized by a bulk nucleation mechanism. The fluoroapatite phase, with appropriate heat treatment, will form as thin needle-like crystals with a high length to diameter aspect ratio.

Apatite–mullite glass-ceramics have demonstrated good mechanical properties, therefore showing promise as a bone substitute material. However, it contains Al_2O_3 and has a high fluoride content, which may inhibit its bioactivity. The responses to implantation of these glass ceramics in vivo have been assessed using a rat femur model (Freeman et al. 2003). In this work, one composition had an apatite stoichiometry (Ca:P = 1.67); three were phosphate (P_2O_5) rich, one calcium rich (CaO), and one phosphate rich composition was tested in the glassy state. Four of the materials exhibited evidence of osseointegration and osteoconduction, while one of the phosphate-rich glass ceramic and the noncrystallized glass showed a marked inflammatory response. It was therefore concluded that crystallization significantly improved the bone tissue response and that the composition with apatite stoichiometry elicited the most favorable response, meriting further research.

A porous apatite–mullite glass ceramic, manufactured via a selective laser sintering (SLS) method, has been evaluated to determine its potential as a bone replacement material. Direct contact and extract assays, used to assess its cytotoxicity, did not show any cytotoxic effects (Goodridge et al. 2007). Unfortunately, there was no evidence of an apatite layer forming on the surface when soaked in SBF for 30 days, suggesting that the material was unlikely to exhibit bioactive behavior in vivo. It is hypothesized that the material was unable to form an apatite layer in SBF due to the fact that this glass-ceramic is highly crystalline and the fluoroapatite crystal phase is relatively stable in SBF, as are the two aluminosilicate crystal phases. There was thus no release of calcium and phosphorus and no formation of silanol groups to trigger apatite deposition. Nevertheless, following implantation in rabbit tibiae for 4 weeks, bone was seen to have grown into the porous structure of the laser sintered parts, and appeared to be very close to, or directly contacting, the material surface indicating that this material is able to osseointegrate, even if it is not bioactive.

Apatite mullite glass ceramics are being studied for use in restorative dentistry, as ceramics for dental crowns (Fathi et al. 2005). Mechanical testing, using biaxial flexural strength (BFS) measurement, showed that the BFS increased as the fluoride content (CaF_2) increased but was still significantly lower than a lithium disilicate glass ceramic. Increasing the CaF_2 content has been shown to increase the chemical solubility in both the glassy and the glass-ceramic stage and the solubility values obtained show that these compositions would be suitable to manufacture cores for dental crowns.

14.5 Fluorocanasite Glass-Ceramics

Ideally, a glass ceramic system for bone substitution and augmentation should be osteoconductive and have good mechanical properties. Furthermore, the parent glass should have a low liquidus so that complex shapes may be cast using the lost wax casting technique. One glass ceramic that potentially meets these requirements is fluorocanasite ($K_2Na_4Ca_5Si_{12}O_{30}F_4$), which has a highly crystalline microstructure consisting of interpenetrating blades that result in a relatively high flexural strength (>300 MPa) and fracture toughness (>5 MPa $m^{0.5}$) (Beall 1991). It also bulk nucleates, and is therefore suitable for casting to net shape. The canasite stoichiometry forms a stable glass requiring only a few percent of excess fluoride to achieve efficient nucleation and it is easy to produce an essentially monophase glass-ceramic. Internal nucleation is achieved through precipitation of CaF_2 crystallites and spherulitic growth of canasite upon these nuclei to give a fine-grained glass-ceramic. Likitvanichkul and Lacourse (1995) showed that the F content of the parent glass is critical for the formation of canasite glass-ceramics. Fluorine is required for both nucleation by CaF_2 and, since it is a constituent of canasite, it must also be available during the growth stage.

An initial study by Corning suggested that the addition of a small percentage of P_2O_5 could induce bioactivity in the canasite system. In P_2O_5-modified compositions, CaF_2 and fluoroapatite act as nucleating agents for canasite, which crystallizes at 750°C. The in vitro bioactivity of a range of modified canasite compositions have been assessed by immersion in SBF (Miller et al. 2002). The composition containing a high concentration of CaO exhibited the formation of a silica-rich surface layer followed by apatite formation on its surface within 3 days of immersion in SBF. A surface apatite layer was detected on the P_2O_5 containing composition as well, but without the involvement of a Si-rich layer due to the crystallization of fluoroapatite in addition to canasite within the glass-ceramic. This indicated that canasite and canasite–fluoroapatite glass ceramics both exhibited in vitro bioactivity, but the mechanisms responsible were different.

Kanchanarat et al. (2007) further assessed the mechanical properties of modified fluorocanasite compositions and reported BFSs of ~250 MPa and fracture toughness of ~2.5 MPa $m^{1/2}$. Assessment of the crystalline microstructure using X-ray diffraction (XRD) and transmission electron microscopy (TEM) showed that in stoichiometric canasite, laths of predominantly frankamenite homogenously nucleate throughout the glass at ~700°C without the presence of a nucleating phase. However, in Na_2O-deficient compositions, CaF_2 particles, forming at 650°C, act as nucleating sites for canasite laths at 700°C. In CaO-rich compositions, CaF_2 particles once again act as nucleating sites but form xonotlite ($Ca_6Si_6O_{17}(OH)_2$) rather than canasite. When these modified compositions were evaluated in vivo using a rabbit femur model, they were shown to be more osteoconductive than commercial fluorocanasite compositions (Bandyopadhyay-Ghosh et al. 2010). A composition containing P_2O_5 was shown to be the most osteoconductive, with evidence of extensive bone tissue contact at 4 and 12 weeks following implantation. Histologically, the mechanism of osteoconduction was similar to that observed with calcium phosphate biomaterials, as micrographs did not show evidence of a surface reaction layer seen in 45S5 Bioglass™.

14.6 Potassium Fluorrichterite Glass-Ceramics

Potassium fluorrichterite (KFR) belongs to the family of amphiboles, which are inosilicates of the general formula $XY_2Z_5(Si, Al, Ti)_8O_{22}(OH, F)_2$. The X represents large ions such as sodium or potassium and this site can be left vacant. The Y can be populated by sodium, calcium, iron (+2), lithium, manganese (+2), aluminum, and/or magnesium and more rarely zinc, nickel, or cobalt. The Z can be filled by ions such as iron (+3), manganese (+3), chromium (+3), aluminum, titanium (+4), iron (+2), lithium, and manganese (+2). Most amphiboles are monoclinic, but members of the Anthophyllite Subgroup are orthorhombic. The rather complex possibilities of the general formula results in a wide variety of minerals in the amphibole group. The amphiboles represent the majority of the double-chained inosilicates. Of these, so far only fluorrichterite has been explored for medical and dental applications.

Potassium fluorrichterite ($KNaCaMg_5Si_8O_{22}F_2$) is a double chain silicate amphibole that has a random acicular microstructure. As a result of its microstructure a fracture follows a tortuous path around rod-like crystals resulting in good mechanical properties. The microstructure consists of fine-grained acicular amphibole crystals with an aspect ratio of 10:1 in a matrix of cristobalite, mica, and residual glass.

The evolution of the various phases in this system was first reported by Beall (1991), followed by Omar et al. (1997) and more recently has been discussed in some detail by Mirsaneh et al. (2002). In the stoichiometric composition, the phase evolution begins with the crystallization of mica at 650°C, followed by diopside at 700°C, and KFR at 750°C. Between 750°C and 900°C, mica and diopside are replaced by KFR and at over 900°C KFR is the only crystalline phase in the system. KFR crystallizes due to a reaction of mica and diopside with the residual glass. Stoichiometric samples, which have been heat treated at 900–950°C had a fine microstructure, whereas samples heat treated to 1,000°C possessed a bimodal distribution of crystals with 10–20 μm rods in a matrix of 0.2–0.3 μm lath crystals. TEM and SEM work done on KFR compositions have shown that the residual glass is silica rich. The effects of magnesium and sodium content on the microstructure of KFR have been investigated by Denry and Holloway (2000, 2002). The mechanical properties of these materials are comparable with that of A–W but are much easier to manufacture. KFR can be cast to shape and then heat treated as it volume crystallizes. This contrast with A–W, which surface crystallizes such that the processing involves the use of powder technology, which is more complex. However, under normal circumstances KFR is not considered to be bioactive.

In the same way that there have been recent attempts to improve the osteoconductive potential of canasite glass ceramics by the addition of P_2O_5, this is now being explored for KFR with some promising results (Bhakta et al. 2010). KFR modified with the addition of CaO has a residual glass phase similar to Bioglass™ and is potentially bioactive. Modification of the stoichiometric composition by the substitution of 5 mol% CaO for MgO suppressed mica crystallization. The microstructure is similar but coarser than the stoichiometric composition and contained more of the residual glass phase. KFR modified by the addition of 2 mol% P_2O_5 to

Table 14.3 Chemical composition of GST, GC5, GP2 (after Bhakta et al. 2010)

Code	SiO_2	Na_2O	K_2O	MgO	CaF_2	CaO	P_2O_5
GST Stoichiometric	53.37	3.33	3.33	33.35	6.62	–	–
GC5	53.37	3.33	3.33	28.35	6.62	5	–
5 mol% CaO subs. for MgO in GST							
GP2	52.26	3.26	3.26	32.66	6.56	–	2
2 mol% P_2O_5 added to GST							

the stoichiometric composition, followed a pattern of crystallization similar to the stoichiometric composition except for the formation of fluorapatite at 800°C and enstatite at 1,000°C. The stoichiometric composition along with experimental compositions with the addition of CaO (GC5) and P_2O_5 (GP2) are presented in Table 14.3.

Further addition of P_2O_5 above 5 mol% to the stoichiometric composition result in severe liquid–liquid phase separation and very poor mechanical properties. The microstructure of GP2 was found to be similar to that of GST, containing small laths (0.2 μm) at 950°C and larger ones (10 μm) at 1,000°C. However, at 1,000°C, some of the larger laths were identified as enstatite using EDS, while some of the smaller laths were identified as fluorapatite using TEM (Mirsaneh et al. 2008). All compositions showed the formation of a calcium phosphate-rich surface layer in SBF an were likely to be osteoconductive in vivo, with GP2 providing the best performance in terms of the combination of rapid formation of the surface layer and superior mechanical properties. In vivo osteoconductivity assessment carried out by implantation of test specimen into healing defects within rabbit femurs indicated that GST, GC5, and GP2 were all osteoconductive in vivo (Bhakta et al. 2011). Therefore, this glass-ceramic system has potential as a load-bearing bioceramic for fabrication of medical devices intended for skeletal tissue repair but this will require further investigation.

References

Beall, G.H.: Chain silicate glass-ceramics. J. Non Cryst. Solids **129**, 163–173 (1991)

Bhakta, S.: Assessment of modified potassium fluorrichterite compositions as load bearing biomaterials in the craniofacial skeleton. PhD thesis, University of Sheffield, UK, 2011

Bhakta, S., Miller, C.A., Brook, I., Van Noort, R., Hatton, P.V.: Prediction of osteoconductive activity of modified potassium fluorrichterite glass-ceramics by immersion in simulated body fluid. J. Mater. Sci Mater Med. **21**, 2979–2988 (2010)

Cerruti, M., Greenspan, D., Powers, K.: Effect of pH and ionic strength on the reactivity of Bioglass 45 S5. Biomaterials **26**, 1665–1674 (2005)

Denry, I., Holloway, J.A.: Effect of magnesium content on the microstructure and crystalline phases of fluoramphibole glass-ceramics. J. Biomed Mater Res (Appl Biomater) **53**, 289–296 (2000)

Denry, I., Holloway, J.A.: Effect of sodium on the crystallization behaviour of fluoramphibole glass-ceramics. J Biomed Mater Res Appl Biomater **63**, 48–52 (2002)

EL-Meliegy, E.M., El-Bassyouni, G.T.: Study of the bioactivity of fluorophlogopite- whitlockite ceramics. Ceram Int **34**(6), 1527–1532 (2008)

Fathi, H., Johnson, A., Van Noort, R., Ward, J.M.: The influence of calcium fluoride on biaxial flexural strength of apatite-mullite glass ceramic materials. Dent Mater. **21**(9), 846–51 (2005)

Fini, M.: Biomaterials for orthopedic surgery in osteoporotic bone. Int. J. Artif. Org. **20**(5), 291–7 (1997)

Franks, K., Abrahams, I., Knowles, J.C.: Development of soluble glasses for biomedical use. Part I: In vitro solubility measurement. J Mater Sci Mater Med. **11**(10), 609–614 (2000)

Freeman, C.O., Brook, I.M., Johnson, A., Hatton, P.V., Hill, R.G., Stanton, K.T.: Crystallization modifies osteoconductivity in an Apatite-Mullite glass-ceramic. J. Mater. Sci Mater. Med. **14**, 985–990 (2003)

Ghosh, B.S., Faria, P.E., Johnson, A., Reaney, I.M., Salata, L.A., Brook, I.M., Hatton, P.V.: Osteoconductivity of modified fluorocanasite glassceramics for bone tissue augmentation and repair. J. Biomed. Mater. Res. A **94**(3), 760–768 (2010)

Goodridge, R.D., Wood, D.J., Ohtsuki, C., Dalgarno, K.W.: Biological evaluation of an apatite-mullite glass-ceramic produced via selective laser sintering. Acta Biomater. **3**(2), 221–31 (2007)

Gross, U.M., Strunz, V.: J. Biomed. Mater. Res **14**, 607 (1980)

Hench, L.L.: Bioceramics: from concept to clinic. J Am Ceram Soc **74**, 1487 (1991)

Hench, L.L.: The story of bioglass. J Mater Sci Mater Med **17**, 967–978 (2006)

Hench, L.L., Clark Jr., A.E., Schaake, H.F.: Int J Non Cryst Sol **8–10**, 837 (1972)

Hill, R.G., Stevens, M.M.: Bioactive glass. US patent 0208428, 2009

Kanchanarat, N., Bandyopadhyay-Ghosh, S., Reaney, I.M., Brook, I.M., Hatton, P.V.: Microstructure and mechanical properties of fluorocanasite glass-ceramics for biomedical applications. J Mater Sci **43**(2), 759–765 (2007)

Knowles, J.C., Franks, K., Abrahams, I.: Investigation of the solubility and ion release in the glass system K_2O-Na_2O-CaO-P_2O_5. Biomaterials **22**, 3091–3096 (2001)

Kokubo, T., Takadama, H.: How useful is SBF in predicting in-vivo bone bioactivity. Biomaterials **27**, 2907–2915 (2006)

Likitvanichkul, S., Lacourse, W.C.: Effect of fluorine content on crystallization of canasite glass-ceramics. J Mater Sci **30**(24), 6151–6155 (1995)

Marikani, A., Maheswaran, A., Premanathan, M., Amalraj, L.: Synthesis and characterization of calcium phosphate based bioactive quaternary P2O5 – CaO – Na2O–K2O glasses. J. Non Cryst. Solids **354**, 3929–3934 (2008)

Miller, C.A., Kokubo, T., Reaney, I.M., Hatton, P.V., James, P.F.: Formation of apatite layers on modified canasite glass-ceramics in simulated body fluid. J. Biomed Mater Res **59**(3), 473–80 (2002)

Miller, C.A.: Crystallisation of canasite/frankamenite based glass-ceramics for bone tissue repair and augmentation, PhD Thesis, University of Sheffield, 2004 (M008947SH)

Mirsaneh, M., Reaney, I.M., James, P.F.: Phase evolution in K-fluorrichterite glass-ceramic. Phys. Chem. Glasses **43C**, 317–320 (2002)

Mirsaneh, M., Reaney, I.M., Hatton, P.V., Bhakta, S., James, P.F.: Effect of P2O5 on the early stage crystallization of K-fluorrichterite glass-ceramics. J. Non Cryst. Solids **354**, 3362–3368 (2008)

Nakamura, T., Yamamuro, T., Higashi, S., Kokubo, T., Itoo, S.: J. Biomed. Mater. Res. **19**, 685 (1985)

Omar, A.A., EL-Shennawi, A.W.A., Hamzawi, E.M.: Effect of isomorphous substitutions on crystallisation of fluorrichterite glasses. Key Eng. Mater **132**, 836–839 (1997)

Piotrowski, G., Hench, L.L., Allen, W.C., Miller, G.J.: J Biomed Mater Res Symp **9**, 47 (1975)

Tirapelli, C., Panzeri, H., Soares, R.G., Peitl, O., Zanotto, E.D.: A novel bioactive glass-ceramic for treating dentin hypersensitivity. Braz Oral Res **24**(4), 381–7 (2010)

Wolfram, H.: Biocompatible and bioactive glass-ceramics – state of the art and new directions. J Non Cryst Solids **219**, 192–197 (1997)

Yamamuro, T.: A/W Glass-Ceramic: Clinical Applications. In: Hench, L.L., Wilson, J. (eds.) Handbook of Bioactive Ceramics: vol. 1: Bioactive Glass and Glass Ceramics. An Introduction to Bioceramics, p. 89. World Scientific, Singapore (1993)

Yoshii, S., Kakutani, Y., Yamamuro, T., Nakamura, T., Kitsugi, T., Oka, M., Kokubo, T., Takagi, M.: Strength of bonding between A-W glass-ceramic and the surface of bone cortex. J. Biomed. Mater. Res. **22A**, 327 (1988)

Zanotto, E.D., Ravagnani, C., Peitl, O., Panzeri, H., Lara, E.H., inventors.: Process and compositions for preparing particulate, bioactive or resorbable Biosilicate for use in the treatment of oral ailments. Patent WO2004/074199, 2004

Index

E. El-Meliegy and R. van Noort, *Glasses and Glass Ceramics for Medical Applications*,
DOI 10.1007/978-1-4614-1228-1,